Entering the Mind of the Tracker

"A deeply moving book, wonderfully written; brings home the tremendous beauty of depth perception of the natural world and the exquisite intelligence and sensitivity of our kin—the Wolves. It opens up the possibility for us to read the world around us through the sense perception born within us, and opens up the potential for us to reinhabit the world." —STEPHEN HARROD BUHNER, AUTHOR OF *THE SECRET TEACHINGS OF PLANTS* AND *ENSOULING LANGUAGE*

"*Entering the Mind of the Tracker* points to the wisdom of asking questions rather than having answers." —PAUL REZENDES, AUTHOR OF *THE WILD WITHIN* AND *TRACKING AND THE ART OF SEEING*

"*Entering the Mind of the Tracker* beautifully demonstrates that outdoor skills are best learned through a deep understanding of environment. Tamarack Song imparts knowledge in a way that is both lyrical and philosophical." —TRISTAN GOOLEY, EXPEDITION LEADER, AUTHOR OF *THE NATURAL NAVIGATOR*, AND THE ONLY LIVING PERSON TO HAVE BOTH FLOWN AND SAILED SOLO ACROSS THE ATLANTIC

"Real stories from the real world, simple and complex at the same time, and well worth pondering!" —BILL McKIBBEN, AUTHOR OF *THE END OF NATURE* AND *EARTH: MAKING A LIFE ON A TOUGH NEW PLANET*

"I couldn't put this book down. Tamarack is not telling us the mere mundanities of tracking—he's showing us a complete communication system that is largely unknown to modern man. Beginning trackers and generalists will love this window into the world of hidden knowledge, and experts will find these stories helpful, insightful, inspiring." —CHRISTOPHER NYERGES, PRIMITIVE SKILLS INSTRUCTOR, AUTHOR OF *HOW TO SURVIVE ANYWHERE*, AND FORMER EDITOR OF *WILDERNESS WAY MAGAZINE*

"Come in from the cold and warm your heart by the fire of tradition. Master tracker and storyteller Tamarack Song shares the stories and wisdom of a life spent in search of the Ancestral Self. It is trailcraft for the soul." —STEVE WATTS, ABORIGINAL STUDIES PROGRAM, SCHIELE MUSEUM OF NATURAL HISTORY; AND PRESIDENT OF THE SOCIETY OF PRIMITIVE TECHNOLOGY

"To track is to live the life of the quarry—mentally, spiritually, and physically. Very few trackers ever reach this level of mastery. *Entering the Mind of the Tracker* will help you discover the salient truths known by those few, like Tamarack, who have shadowed all living things. Here is a window to the beautiful and foundational knowledge provided by a lifetime of tutelage at the feet of Mother Nature." —Ty Cunningham, founder and tracking historian of the International Society of Professional Trackers

"Tracking has become a left-brained skill, involving ruler, track analysis, and GPS to know an animal from its tracks. In *Entering the Mind of the Tracker*, Tamarack Song offers an intuitive, Zen-like alternative, suggesting that we don't need to learn to track any more than a Wolf needs to be reminded that he is a hunter. Tracking is in our nature. Whether you are an experienced tracker seeking to improve your ability or a novice intimidated by the left-brained science of conventional tracking, this book opens up exciting opportunities to connect with the story of the land." —Thomas J. Elpel, author of *Botany in a Day* and founder of Hollowtop Outdoor Primitive School

"*Entering the Mind of the Tracker* is a marvelous book written by a master storyteller and tracker. Through its powerful and poignant stories you will feel absorbed in the world and spirit of Nature." —Joseph Cornell, founder of the Sharing Nature Foundation and author of *Sharing Nature with Children* and *John Muir: My Life with Nature*

"When a human being has passion for wild places, / And pauses, comprehending the spaces around, / Then the tiny notice of a bent twig graces / All the story inside that certain spot of ground. / Such is the inner passion of Tamarack Song, / His stories reflecting the seeing parts of his days / That are so large and informing to us who long/For the revelation of observation's ways." —Larry Dean Olsen, author of *Outdoor Survival Skills*, progenitor of wilderness therapy, and founder of the Stone Age skills movement

"Brilliant! So fresh, so enticing, even seasoned trackers will be blown wide open, the novice will be jump-started years ahead, and teachers and guides will rejoice! Tamarack Song melds ancient wisdom with modern knowledge and offers us a fun, full, fantastic learning experience." —Robin Blankenship, founder of Earth Knack Primitive Skills School and author of *Earth Knack: Stone Age Skills for the 21st Century*

Entering the Mind of the Tracker

of the Tracker

Native Practices *for* Developing
Intuitive Consciousness
and Discovering Hidden Nature

Tamarack Song

Bear & Company
Rochester, Vermont • Toronto, Canada

Bear & Company
One Park Street
Rochester, Vermont 05767
www.BearandCompanyBooks.com

Bear & Company is a division of Inner Traditions International

Library of Congress Cataloging-in-Publication Data

Song, Tamarack, 1948–
  Entering the mind of the tracker : native practices for developing intuitive consciousness and discovering hidden nature / Tamarack Song.
    p. cm.
  Includes index.
  ISBN 978-1-59143-160-2 (pbk.) — ISBN 978-1-59143-827-4 (e-book)
  1. Tracking and trailing. 2. Animal tracks. 3. Indians—Hunting. I. Title.
  SK282.S66 2013
  591.47'9—dc23

                          2012036149

Printed and bound in the United States

10 9 8 7 6 5

Text design by Brian Boyton and text layout by Virginia Scott Bowman
This book was typeset in Garamond Premier Pro and Myriad Pro with News Gothic and Bodega Serif used as display typefaces

Field sketches by Mark Webster

All appendix illustrations and data are based on the author's specimen collections, field notes, and photos.

To send correspondence to the author of this book, mail a first-class letter to the author c/o Inner Traditions • Bear & Company, One Park Street, Rochester, VT 05767, and we will forward the communication, or visit the author's websites at **www.tamaracksong.org**, **www.teachingdrum.org**, and **www.brotherwolffoundation.org**.

The chapters of my life are tales of Wolf tracks across my days and dreams. I am grateful for this opportunity to share the story of what I have learned from Wolf and his animal kin, and to dedicate this book to the future of Wolf and human living again in harmony.

All author proceeds from this book go to support the creation of the Brother Wolf Foundation, a nonprofit sanctuary for Timber Wolves rescued from puppy mills and backyard pens. Open to the public, the sanctuary will reintroduce people to the once-respectful and mutually beneficial Wolf-human relationship. With Wolf's inspiration and example, a new generation can awaken to their innate tracking abilities and learn the ways of living in balance, just as our hunter-gatherer ancestors did. For more information, go to www.brotherwolffoundation.org.

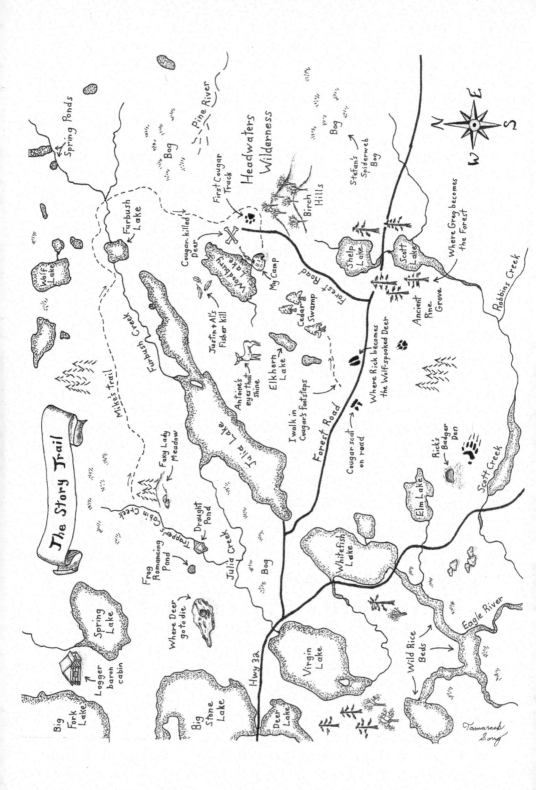

Contents

Foreword

My name is Bryan Nez, and I have been a tracker in the deserts and mountains of the Southwest for most of my life. In early 2010 I received a mysterious package in the mail. In it there was a manuscript titled *In the Shadow of Wolf* [this book's original title]. Considering my time on the U.S. Customs unit called the "Shadow Wolves," I was a bit interested, so I opened it up. I then couldn't put down. It brought me back to the way I learned tracking from my father and grandfather, who taught it the Old Way.

I teach tracking all over the world now, yet I don't stray far from my roots. I was born in Winslow, Arizona, and then started off life in a little town called McNary, near the New Mexico border. Since I am a full-blooded Navajo Indian, my parents wanted me to be near my grandfather on the Navajo Reservation, so we moved back when I was very young. There I learned about the environment we lived in and about how to be with it. It was here that I began to learn how to track. I assume that a fair number of people reading this book are looking for clues on how to be a better tracker, and in many ways it starts here, with a book just like this. But first, I have a little scenario for you.

Imagine yourself stepping into a warehouse and from the front door to the back wall there is nothing but file cabinets. That's your life—at the far end is your birth and up at the front door, that's the present. You see yourself starting at the front and going through the files, all the way to the back of the building. There, after the last cabinet, is a small door on the wall with a knob and hinges that are all rusty. You open it and it creaks—it's stiff and heavy. Inside it is dark, and there is dust on the ground and cobwebs hanging. At the far end of the room, in a corner, you see a little bundle. You walk over to it, pick it up, and bring it out. Brush off the cobwebs and dust, open it up, and you have all your tracking and survival techniques from the Old Days—it's all right there. Find it, and I'll tell you this: within a day, you'll be tracking.

You'll keep getting better with practice, but what happens is that both seasoned trackers and new trackers forget they have this bundle. Time and again—even when I pick up this book—I am reminded of something that is easy to forget nowadays, which is those things that keep us connected to our bundle: its story, community, history, and tradition that are so important in the Old Ways. Many of my students confess before we start tracking: "I've never done this before," "I'm not a tracker," or "I don't know how to track." I tell them, "It's in you already; it's just that your mind is clouded. You've got to open and relax, and your bundle—your tracking mind—is there." It's not just staying away from computer gizmos. Let me illustrate by telling a little more of my story.

My training began as a game. Like many on the reservation, I attended boarding school, in a place called Leupp, Arizona. It was very strict, military-like training—you had to march to wherever you were going. I had long hair, and they cut it down like a military crew cut. We were told not to speak our own language, to speak English only. If you got caught speaking Navajo, you'd get a "whupping" for it.

On weekends the parents would come and check you out of school; but if the parents didn't show up, the kids would be stranded at the school over the weekend. Some of us kids would try to run away because we were homesick, and sometimes home was seventy miles away.

I was one of the kids that tried to run away. The older kids were trackers, so the teachers would send them out to get us. They'd come after us and find us every time. We'd sit around out there and after a while they'd say, "Okay, we'll give you an hour; you take off and we'll see if we can find you again. Go!" And off we went. We'd try to sneak around and backtrack, do all sorts of stuff to try to get away from them, but we never did. When I grew up a little, I became one of the trackers, and I'd play the same game with the younger ones.

That was some of my training. At home on the reservation, my mother and grandfather got me going on tracking, counter tracking, and survival—they taught me what vegetation was edible and what was for medicine—and what was poisonous. I learned to hunt, cook, and skin wild animals. We had a lot of night training, a lot of war games between relatives. Story was a good part of the training: my grandfather told us stories of the neighboring tribe coming into our territory and acting badly. That was his side of the story, and I think he told us things like that to keep us sharp and out of trouble.

Stories and tradition tell us what to do, what not to do. I'd tell any new tracker to begin with tradition and history and stories just like in this book. If they're going to track, don't first go out and find tracks. In the Old Days, things were passed on by word of mouth. There were no written languages, so everything was verbal—it's how the ceremonies, the songs, and the traditions were kept intact and passed on from one generation to the next. This is how I teach, and this is how we learn. I tell a young tracker stories about how turkey got his white tail or how bear got his feet switched around, which is why he walks with his toes turned in. Even little stories have power—go out and find stories—they teach so much more than any tracking manual out there. That's why this book of stories *is* a manual on tracking, and one I value.

Tracking is story, and all tracks tell one. Once you get good at listening to stories, you'll be better able to hear the stories of the tracks. Stories teach us how to be thankful for the animals and for the land that we live on. I believe this is a first step for any tracker. Talk and give thanks to the creator. Sit down outside and listen to the trees, to the wind blowing through the branches and the leaves swishing. Then go talk to the elders—listen to the ones who knew a different life. Ask them about their experiences, and they will teach you the ways that you might not know yet. And then go to another tracker, learn the best of his stuff; and go to another tracker and learn the best of his stuff. Combine those skills and make them your own.

If you're looking for tracks, you have to situate yourself to listen. You have to listen to what the footprint is telling you, and that's not going to be heard by staring straight down into the track. Tamarack's book reminds us to step back and listen to the story of the being who made the track. Everything leaves sign, no matter how good you are. If you're a cougar, you leave a track; if you're a man, you leave a track. Even the wind leaves a track.

But we forget. Often I tell my students that I'm still a student too, and will be until the day I die. Out in the field I'll ask them, "What do you see?" and they'll mention something that I'd forgotten. It's a cycle—like the seasons, everything goes in a circle. When you teach someone something, they'll show you something that you had forgotten, and it returns to you. This book is like that—the stories are reminders of basics that we need to stay in touch with. The stories are to pass down, to tell around a fire or a meal, or around a track deep in the woods. Read these stories and share them, and tell your own stories and share them often. They will remind you. When you teach or tell a story, you've unknowingly recorded something for those listening, even though it's not written. It stays with them, and they'll come back and tell you, in one form or another, and you'll remember again.

I read this book cover-to-cover, and I will go back to it again. It is a bible of sorts—the best tracking manual there is—it is all there. And I value it because it is the Old Way—it holds what I learned from my dad and grandfather. We are quickly forgetting the Old Ways, not just in tracking, but in how we are with the world. I hope you find what I have found here, and I hope you continue to share it—it's something that we need to keep alive.

BRYAN NEZ,
FORMER SHADOW WOLF

Bryan Nez is a retired senior member of the legendary Shadow Wolves, the elite U.S. Immigration and Customs Service unit that apprehends drug smugglers crossing the U.S./Mexican border. Based on the Tohono O'odham Indian reservation in southern Arizona, the unit is comprised solely of American Indian men and women who use traditional tracking skills handed down from generation to generation. Bryan now trains the tactical tracking units of many nations around the world.

Tracking in the Shadow of Wolf

She feels it deep in her bones: old, old memories she has never lived, yet she knows them like she has. She hears it: a guiding voice that is already inside her, while at the same time it comes from all around like an echo in a box canyon. Not only with her ears does she listen, but with all of her senses. And she sees it, not with her eyes, but throughout her body, in every muscle and nerve. It gets hold of her, enticing her to go quickly this way and then that, to be bold and then cautious. All the while the exhilaration—the tingling of her every fiber—melds her seamlessly with her surroundings. She and her prey are one—she was clearly made to do nothing else but track.

This is what a Wolf* experiences on the hunt, as I observed it while living with them. It is no different from what you or I would feel when on the trail of an animal, because we are just like Wolf—we are born trackers.

We rely on our tracking ability throughout the day. In fact, it is so vital to our existence that we would be literally lost without it. Whether we are finding our way across town, tracking down a set of keys, keeping track of the kids, or simply shopping, we are tracking. But our ability does not stop there—it extends to the qualitative aspects of life, such as planning a vacation, looking for a mate, wondering why we feel the way we do, and sensing what others are thinking.

We do all this and more, even though we may never have taken a tracking course. We inherited the talent from our Paleolithic ancestors, who were already hunting cooperatively 400,000 years ago† and developing a tracking ability unique to the animal world. They could envision scenarios, whereas other animals had to rely primarily on their senses and intuition. Over the generations, human track-

---

*Out of respect for my relationship with Wolf, along with the other animals and plants who have been my family and guides, I capitalize their names in this text, and I refer to them as *he* and *she* rather than *it*.

†Stiner, Mary C., Ran Barkai, and Avi Gopher, "Cooperative hunting and meat sharing 400–200 kya at Qesem Cave, Israel," *Proceedings of the National Academy of Sciences* 106, no. 32 (2009), doi: 10.1073/pnas.0900564106.

ing evolved into the highly refined, inheritable trait that guides us through our days.

As with all innate abilities, our tracking aptitude improves with practice. We had to find our way around town a number of times before we could do it with confidence, and the same is true of trailing animals. This book is a tool that will guide and inspire you to gain the experience to improve your tracking ability. You will increase your working knowledge of animal behavior and all of nature. It is my hope that the tracker inside you will awaken while you walk alongside the characters in the story. And then the next time you wander into the farther places, you might see and hear in a deeper way—a way that draws you into the song of the track.

But this is not just about finding animals or lost keys. As we become better trackers, life in general transforms into a healthier, richer experience. Tracking becomes a metaphor for conscious living as we apply what we have learned on the trail. We become more connected with the life around us and we grow more in touch with our inner life—our passions and feelings. We learn to better express ourselves and listen. We are more able to meet our needs and help others. We discover how to embrace the Zen—the essence—of life.

To become an effective tracker is to become the animal we are tracking. If we can feel the pull of her needs and the strain of her fears, we will know why she does what she does and where we can find her. When we have this kind of intimate connection with the life around us, we are going to be caring and responsible earth stewards. The purpose of this book is to help us become the animal we track and make her world ours.

To become the animal, we first need to be hungry. An animal hunts to eat, and hunger is her motivation. A full belly numbs a Wolf's senses and puts her to sleep, while hunger super-sensitizes her so she can feel, see, and hear things she might otherwise miss. We are designed the same way: our ancestors became the animal and tracked for the same reason as Wolf.

If we are going to learn well, our drive must also be hunger. There are many kinds of hunger, so our motivations will not all be the same. Except for one—failure. Success dulls, failure teaches. Expect your progress to be slow at first, and you may get confused and frustrated. Invite it and be thankful for it, because it means that you are learning.

Keep in mind that you are in good company when you fail. As experienced as a Wolf might be, he is hardly always successful. He and his pack might chase ten Moose before they take one. The many misses gift the pack with what native Elders tell me is *the teaching trail*. It doesn't take us to the animal, but it gives us the experience we need to learn the way. At the same time, the prey gets the trial they need to become better survivors. The weak, slow, and dull go to the skilled tracker, which keeps the prey strong and bright.

I was fortunate to have learned tracking directly from Wolves, by training with them and listening to their stories. As a kid, I spent all my free time in the woods and swamps surrounding the small farming town where I grew up. To me, Wolf was the mystic symbol of the wilder places beyond. I read, dreamt, and fantasized about Wolves tracking their prey and raising their pups. I wanted to live with them. When I wound through the understory, I was a Wolf; when I napped in the sun, I was a Wolf. I stalked on soft paws and scanned the forest through sharp amber eyes. When I saw a hillside den, it was the home of my pack. It didn't matter that there were no longer any Wolves within hundreds of miles; Wolf's spirit was my spirit.

When I turned twenty-one, I camped for a few days on a high stream bank in a northern Wisconsin wilderness. One night I heard a lone howl. My gut tightened, I stopped breathing, and my eyes strained to make something out. I wanted to know what direction it came from and if there was more than one. At the same time, I wondered if I wanted it to be a howl so bad that I was practicing some creative hearing with a far-off Owl's hoot.

The next morning, I tracked him—or at least I convinced myself that I did—and I found what I thought could be Wolf scat. Nearby were the remains of a long-dead Deer. It had to be one of his kills, so I made sure to find all the supportive sign I needed. A few hairs that were probably from the Deer's belly became the Wolf's, and the sap-stained bark at the base of a nearby Tree turned into his urine scent marking. I don't know what I could have done to stray any further from the tracker's first directive, "Be as a question."

Shortly after that experience, a friend came by to tell me he had stopped in at a little Northwoods roadside zoo that had a pair of Wolves who had just given birth to pups. At first light the next morning, I was on the road and heading north.

Each pen held a well-known Northwoods animal: Porcupine, Raccoon, Beaver, Fox, Red and Gray Squirrel, White-Tailed Deer, and Black Bear, along with a huge Snapping Turtle and a ghostly looking albino Deer. The centerpiece of the menagerie was a pair of Wolves.

But where were the pups? "In the house," said the proprietor. The place wasn't busy and he was an obliging person, so I was soon looking down into the bathtub at four eleven-day-old pups with eyes still not open. He said they had just been taken from the pen the night before.

I asked how much he would take for a pup.

"Two hundred dollars," he replied. "But if you want one, you'll have to take it now. We're leaving for the weekend and don't have anyone to take care of them."

"Will you take four hundred dollars for the litter?"

One pup died right away, for no obvious reason, while the others thrived on regular bottle feedings and all the attention they could handle. Don, a Mahican Indian friend, along with his grandmother, a tribal Elder, helped name them. The female, who grew up to be lanky and inquisitive, became Simbut Meaxtkaoin Mahican, or *Silver Wolf.* She had a determined way of being that resonated with mine—we were soul mates. The middle pup, simply called *Wolfie,* was Simbut's agile male counterpart and future mate. I could already see within the first few days by his cleverness and attention to detail that he was destined to be the pack leader. Deshum Nashak, or *Earth Thunderer,* was half-again bigger than his brother. His size commanded respect, but it didn't take long for anyone to see that he was just a big, gentle teddy bear.

Very soon after their eyes opened, they began the ritual of forming a pack. Being their parent and favorite wrestling buddy, I was included in the process. But not automatically—I had to work out my role as a scout, nurturer, or leader, just as each of the Wolves did. Through eating, tussling, running, and napping together, we came to know each other's qualities and temperaments. Along the way, I learned their complex posture, facial gesture, and vocalization-based language. At first I was confused with what I took to be irrational behaviors, and in time I learned that even the simplest and most subtle of changes could mean something: a blank-looking stare said to back off because something important caught their attention, a slight tightening of the shoulder ruff showed a hint of fear, and a faint loosening of the ruff signaled budding anger. This turned out to be important training for catching the nuances of sign that I could easily overlook when on the trail of an animal.

With Wolves, there is no simple "hello" and "goodbye." They have traditional greeting and parting protocols that took me time to learn and get used to. Joining them usually involved a high-energy exchange of shoves, soft bites, and licks. Before they'd let me leave, I'd have to help them wind down by pinning each one to the ground by the neck and rubbing his belly. But I had to be careful—they liked it so much that I risked getting peed on.

It took me a while to get used to their solid sense of presence and unwavering sincerity. They were upfront with their feelings, which quickly earned my trust and respect. At the same time, my relationship with them caused me to realize how insincere I was in many of my human relationships. Even with the Wolves, I would sometimes out of habit fake being fully present and involved. Only they saw straight through me and responded instead to what I was really feeling. If I drifted off into some fantasy, they would either take advantage of me by nipping and pouncing, or they would ignore me. There was no in between: it was the only way they knew, and the only way they would have me. It forced me to be emotionally honest with them, and in turn with myself.

Sometimes I appreciated the guidance, and sometimes I would react. Who were they to turn their backs on me? I didn't like being called on my stuff and not being accepted for the way I was. Maybe I was insincere now and then, but what else could anyone have expected? I told myself they just didn't understand the emotional charades we humans had to play to get through a day.

At the time, humans took up much of my day, but Wolves filled my heart. Only the Wolves didn't tolerate my whining about work or housemates any more than they put up with my lack of presence. By insisting on sincerity, they had me realize that my struggle was not with other people, but with myself. In my finer moments—when my ego wasn't too bruised—I could see that many of the people I judged were seeking the same connection with the essence of life that I was. In running from my kind, I was only running from myself and into a lonely corner.

Two years later, they had their first pups—three beautiful charcoal-colored balls of fur—in a den Simbut dug under a downed Tree. Watching her try to reject the doghouse-style den I made for her helped me realize the prominent role instinct plays in an animal's life. At first I was anxious about crawling into the den, but Simbut didn't show me even a slight raise of the lip.

With forming a pack and developing hunting skills, Wolfie, Simbut, and Deshum Nashak were making it possible for their descendants to someday re-inhabit their ancestral land. I knew we could progress even more with the pups learning from the pack and growing up on wild game. As soon as they could run, we practiced the skills of the hunt together by playing tag, hide-and-seek, and stalk-and-ambush. Often we'd pair up and read each other's cues to coordinate our chases or find whoever had hidden. By watching them exercise their ability to hear the song of the track, I got more in touch with my own intuitive ability.

Together we discovered and developed our tracking intelligences, which showed how innate they were, by how natural they came to feel. It was as though our young minds had old memories guiding each step on the trail and showing us how to work together as a pack. We were each proficient at a different aspect of the stalk and the chase, and still we worked synchronistically, like the organs in an organism. And it wasn't just us. Blue Jay's warning call diverted the attention of those we stalked, Tree's shadows disguised us, Wind took our scent away. The tracker, I learned, was not any of us alone, but all of us together.

The bigger the pups grew, the more intently they studied each other's moves, and the more complex their games of chase and tag became. The cleverer one got, the faster the other figured it out. They got better and better at anticipating each other's moves—a skill I call *shadowing*—to the point where I feared there wasn't going to be any sport left in the game.

I should have known that reaching the top of one hill only gives a view of the next one. Their shadowing soon took on another level of complexity, with the pursued one skipping her next expected move to throw the pursuer off. Not to be outdone, the pursuer would appear to abandon the chase, only to cut over and intercept the pursued where he thought he was going to be in the clear. A favorite of theirs was to tightly shadow me and then veer off. As soon as I no longer felt anyone at my heels, I knew he'd be waiting right where my move to lose them was going to take me. But it was too late for me to execute a counter-move and save face. I was a Mouse being played with by a Cat.

If those pups played our team sports, they'd have been champions.

Even though someone would occasionally get outsmarted, he seldom gave up. And he always acted surprised when his moves didn't work. I saw it as an example of how success in the hunt—and success in surviving if you are the one being hunted—takes a strong belief in self.

I guess I didn't have the level of self-confidence they did, because I'd sometimes get discouraged and take a sideline seat after getting outsmarted a few times in a row. My excuse was that I was used to playing with Sled Dogs, who had different—easier—game rules. At first I found the Wolf pups to be unpredictable, and that is because I expected them to act like Dog pups. Wolves seemed more catlike than doglike, with minds of their own and a sense of presence that made them appear aloof. They wouldn't easily submit like a Dog, and I imagined they'd be hard to train to any doglike regimen. It didn't take me long to realize I was playing with a truly wild creature, which taught me much about the ways of wild animals in general.

The pups played like the game was everything. They made wrong moves and went tumbling head over heels, only to pop back up and hardly miss a beat. When I wiped out, I'd just lay there laughing at how silly I must have looked. And oh, did I pay for it! They were all over me with dirty paws and slobbery tongues, happy to capitalize on my quirky human behavior.

Those pups could be completely consumed in play and still switch gears in a flash. A sudden wind or strange sound could have them freeze with ears perked. Sometimes whatever it was would dampen their spirits for play and they'd wander off, take naps, or stare off into space. These seemingly spontaneous acts got me used to animals' free-form thinking and helped me look differently at confusing signs.

At first when I was abandoned in the middle of play, I took it personally. I'd never think of turning my back on someone without an explanation. It took longer than I care to admit for me to realize they treated each other the same way they treated me. I eventually saw they didn't need to state the obvious—they were responding to a greater voice: the song of the track. Rather than snubbing anyone, they were caring for each other—me included—by giving their attention to the bigger picture. In order to become a consistently effective tracker, I needed to learn that perspective guides focus.

One day when the pups were about half grown, we were out in a field hunting Meadow Voles. "Quit playing," I scolded them jokingly. Right then it hit me: *the hunt is a game,* so playing games is serious business because it is training for the hunt. No wonder they gave their all to play, and no wonder that even in the heat of sport they kept attuned to the song of the track.

Several years passed, the pack had pups every spring, and my idyllic apprenticeship continued. Along the way, I got involved in educational efforts to replace the Big Bad Wolf image many held with that of the highly social, compassionate creature he truly was. With such a photogenic and controversial topic, the media was right there to assist. They created such interest that on weekends an endless parade of cars would creep down our normally quiet country road. Kids' faces pressed against windows in hopes of snatching a glimpse of the fabled creatures.

And then there were those who believed the only reason the Wolf existed was to be exterminated. An armed contingent stopped by one day to talk about the subject. Alcohol gave them the courage, and unfortunately it also transformed their fear of Wolves to anger. The authorities showed up before a shot was fired, yet the gentlemen accomplished their mission. Declared a public danger because of the passions they unleashed, the pack was taken into custody and split up, with each member sentenced to life behind bars.

I imagined traveling around to the various zoos to visit them. There I stood with the grandparents and kids, peering through a steel grid at hollow shells with lifeless eyes. Even if I could stomach it, what would I say? What if they went crazy and attacked the fence, trying to reach me to plant their paws on my chest and push me over, as they always did? Or what if they didn't respond at all—what if they were too far gone to know or care? I needed to walk on, to never see them again.

Thirty years later, I stood gaping at tracks nearly the size of my hand lacing the edge of the snowbound logging road leading to the trailhead of my wilderness camp. First glance showed one Wolf, but the song of the trail hinted at more. I recalled my old pack stepping in each other's tracks to save energy in deep snow, so I took another look and saw the occasional duplicate print that indicated a pair. What a feast—the first evidence of Wolves in the Northwoods in many decades! I closed my eyes and luxuriated in a flood of memories.

No matter where I looked over the next turn of the seasons, I found no Wolf sign. That changed late in the summer when my two youngest children and I went paddling on one of our favorite wilderness rivers. At sunset, everything quieted. Not a Bird sang and the river's surface was as smoothly polished as the silvery sky. We quit paddling and sat still as the lily pads.

Over the valley drifted a lone howl, echoed by another. Immediately everything vanished but those two voices. Their unsteadiness, their pitch and timbre, their shy volume, took over my consciousness. If I were standing, I would have buckled. This was not just an adult pair who wandered in from somewhere, nor were they transplants who might or might not hang around. And neither were they fragile pups who may not survive. No, these were frisky adolescents, born and raised here—true natives! I savored a moment of contentment.

On a cloudy, unseasonably warm afternoon during the next winter, Justin, a student in my year-long wilderness skills course, asked me to take a look at some scat he came across in a small clearing just a stone's throw from my wigwam. His sparse verbiage told me this was no every-day scat, and so did my intuition. I didn't need to go identify the scat, but trusting notions is not a characteristic of a dependable tracker. Nor of someone who honors his most important relationships, which this sign signified. Still, my urge to run and tell Lety, my mate, only reinforced my first impression.

What a scat! It was plump, richly peppered with bone chips, and finely laced with auburn-colored Deer hair. The plug, the first part to come out, was as well-shaped as the working end of a good fire-by-friction drill. And the cone, the last part, tapered artfully down to just a few hairs. This guy was big and healthy and eating well, and he looked to be in the prime of his life, probably three to seven years of age. The secondary sign, which included the scat's prominent placement, the bold scrape caused by the Wolf kicking snow when finished, and the high-placed urine scent mark on the nearest upwind Tree, told me he was likely a pack's alpha male.

After Justin left, I sat in front of the wigwam and listened for the song of the Wolf's track. At some point—I seldom remember when it happens—my sense of time left, and I watched him come down from the north and skirt the bog in front of my lodge. He was alone, and he moved with purpose. With all the human sign around, I would expect him to give this area wide berth; however, I looked directly at him as he passed close by, and he gave me no notice.

At the far end of the clearing, he sniffed the base of several Trees, marked one, and squatted facing the direction from which he came. Having finished, he sampled the air, then his scat, to imprint his own scent. Satisfied, he went back to his scent Tree, smelled it again, and raised his leg high to re-mark it—heavily this time. Without looking back, he continued on south.

Following communication protocol, I went down to sniff his scent post and mark an adjacent Tree. When he came back this way, my marking would tell him who I was and that I recognized and accepted his presence.

Right after the next snow, I tracked two hunters, one a bit smaller than the other. They came out of the wilderness, ran down a Deer, feasted, and went back. They returned twice more to feed on the carcass, which lay within sight of the camp trailhead. Again they came so close, and again they left sign so conspicuous it seemed intentional. Was a relationship evolving? Even though I was beaming, I knew from the throes of failure that the most essential Wolf habitat lay in my heart. When Wolf could feel at home there, he could in the woods around me. If I wanted to run with him, I had to be as free as him. I needed to reckon with how his way clashed with the way I was raised, which was to plant rather than hunt and stay on the trail rather than follow the feeling.

The clash of the old ways with the new was nowhere as blatant as with the neighboring Ojibwe people. Their Cougar, Martin, Sturgeon, and Wolf clan animals were wiped out, but still the people continued with their traditional ceremonies honoring the animals as though they were there. Why didn't they bury their shattered past, as I had?

I took my question to the Elders, who told me about the long-ago time when Wolf and human hunted together. Through their kinship, they could see their kinship with all life. Still, the time came for Wolf and human to part. A prophecy was given.

> *You shall live as two—you have separate journeys to walk. Still, your spirits will remain as one, and as goes the fate of one, so will go the fate of the other. Some of you will come to fear and reject each other, and you will know each other only as images. Yet remember: if you can continue honoring your relationship during this time, you will again run and hunt together. Watch and learn from each other, so that you will never again have to be apart.*

So the Wolf clan people would not forget the reason for the separation, they called Wolf *Ma'ingan: He-Who-Makes-Strange.* The European invader's predator extermination policies only accelerated the prophecy's unfolding.

Deep down, I understood the Elders' words. In the past, I had been able to hear only part of Wolf's song—the Elders' voice was missing—so I mistakenly thought the time had come for the rejoining of Wolf and human. Neither was ready, yet I wanted it so much. I made myself a trailblazer, and I only ended up fighting destiny and causing misery. Or maybe it was simpler than that, and mine was just the fate of the fool. Another case of a dreamer with delusions. Then again, this could have been another wrinkle in my unfulfilled childhood vision of a free-running life.

The prophecy grounded me. I realized I was not tracking, but forcing the trail to read what I wanted. Numbing out on self-pity was no longer an option—I now knew I was part of something beyond myself, something I was to serve. The entire experience was a key teaching in how to embrace the Zen of the track.

Still, I could not find peace with the way Wolf and I were made strange—until one stone-still night on the high bank of a fog-blanketed river. Moonlight gave the waterway the luminance of a mystic trail winding up a shadowed valley. One of my camp mates, a woman with long, dark hair, started to play a drum. With each beat, firelight rippled and shattered on the drumhead. A shape slowly took form and I strained to make it out . . . eyes . . . a muzzle . . . a heavy muzzle. It was him, Deshum Nashak!

The beat of the drum became his pulse as he took full form. While struggling to get my watering eyes to focus, I fought back a tidal wave of feelings—I needed to stay present. Deshum Nashak helped by acting as though we had just been together yesterday. As usual, he wanted to get right to his favorite game: the staredown. There wasn't a Wolf I couldn't beat, which was no big feat if you knew Wolves never focused long on any one thing for fear of losing perspective.

I was a slow learner back then—I used the word *never*, which does not exist in the vocabulary of the ever-questioning tracker. Deshum Nashak held my focus as I was drawn through the drum-

head. He bounded toward me from across the lush meadow, jumping high to keep me in sight. His nimble form and puppy-like face belied the fact that he weighed nearly what I did. I jumped the meandering stream and we met in a reverie of wild leaps, pinned each other down by the throat, and exposed our bellies to show trust. Like sprightly Deer, we jumped the stream again and again for the sheer joy of it. The entire time he talked to me in the intuitive language known to all natural life. I am well, he said, as is our pack. You can see that Mother Earth is in her primal splendor here. The Deer are fat, and we hunt them together with the humans, as was our way long ago.

Drum's heartbeat left my chest and I returned to the fire. The drumhead showed only cavorting fire shadows. When the fire faded to embers, I slipped back into the forest shadows and knelt with nothing to say, nothing to offer but a small gift of tears. They fell before me on the ancient trail that still remembered the feel of my Wolf family's ancestors' feet.

When he first looked into the "fierce green fire" of a Wolf's eyes, naturalist Aldo Leopold found something known only to her and to the mountain.* In the depths of those wild eyes, I found the essence of both Wolf and human. I could see we were born to run and hunt side by side, as when we lived in balance with all life. I saw neither a mythic symbol of some idyllic wilderness nor a blood-lusting varmint. Instead, the brightness of those eyes guided my rebirth as an intuitive tracker. But it was not just Wolf, nor did it have to be him. In the upcoming story, you will see how I and others learned from Spider and Grouse and even family Dogs and Cats. All of our animal Relations are wise and highly intuitive, and they wait to guide us.

I am now in my elder years, and I walk as both a blue-eyed human with white hair and a blue-eyed white Wolf. I see not as one or the other, but as a creature of the forest. I stalk quarry not as me, but as a hunter in kinship with all hunters. I tell a story not from my memory but from a universal spring effusing the common experience of killing and feasting and loving and grieving. I strive to be as detached from identity as the chased Deer, with the last thing on her mind being the name of her pursuer. All that matters is the terror of the hunt and the lust for another tomorrow.

The story about learning the skills of the hunt, which I'm now going to tell, took place in the Northwoods of Wisconsin. I've lived so close to this land of pristine lakes, soaring pines, and velvet-carpeted bogs that I feel like it birthed me. When I was a young man, I would be up every day with first light and out paddling a new stream to a hidden lake or following a trail to unravel some mystery beneath the dark cedars. I had no job but this, and my only family was the Otters I swam with and the Owls who came to talk with me high in the hemlocks.

Twenty-five years ago on the fragrant needle-carpeted shore of

---

*Leopold, Aldo, *A Sand County Almanac* (New York: Oxford University Press, 1949), 130.

a quiet lake, I built a wigwam of birch bark sewn with spruce root. Occasionally someone who heard about me would come to visit, and I would take them along on my explorations. They would want to learn what I was doing, which was not just tracking and foraging but true immersion in the wild as I learned it from Wolf.

More and more people came. To accommodate them, I started the Teaching Drum Outdoor School and began offering weeklong courses. The heart of Wolf was the heart of the school. Our classroom was the wilderness and the curriculum was whatever the trail laid before us. In time the course grew to a year in length, with the students living in the lakeside camp I built. The course became a full-time wilderness experience with no books, phones, or town trips to distract us. We foraged much of our food and I taught awareness, communication, and hunting skills the same way the pack taught them to the pups and me: through example and experience. Sometimes a pack member would give me a hand with the obvious, and at other times I was left to flounder—and I did the same with the students. All I asked of them was what the pack asked of me: that I be willing.

Students, staff, and friends lace their way through the story, with Wolf as their invisible, ever-present guide. Most of the chapters are transcribed from live storytelling, to capture the spirit of a one-on-one sharing. The story reflects what I have seen and learned, and even more, it shows how much I have yet to learn. Therein lies the yearning—the thrill—to go and again and again become the animal and move like a shadow. I feel like I walk in the footsteps of the old Lakota grandmother Uncheedah, who said, "When you see a new trail or a footprint you do not know, follow it to the point of knowing."*

I'll start the traditional way my Ojibwe Elders have taught me to begin a storytelling. This person of the Owl clan is called "Tamarack

---

*Eastman, Charles Alexander, and Michael Oren Fitzgerald, *The Essential Charles Eastman (Ohiyesa): Light on the Indian World* (Bloomington, Ind.: World Wisdom, 2007), 165.

Song," and he brings this story from *Gabe Nishnajida* (Ojibwe for Camp Where the Old Way Returns) in the Snowmelting Moon of the Late Snows Winter.

Now imagine we are sitting around a campfire watching the sun set over the lake. The hum of the high Pines mingles with Owl's first tentative *who-who-a-who*. Soon the story will turn the curling smoke into stalking animals and pensive trackers. Listen well, because the best trackers were once the best listeners. So turn the page and we'll learn from Wolf and the other animals as we join them in the chase and the kill. Leave preconceptions behind and together we will go where one question leads to another, and then another and another.

1
Sweet Fern Rendezvous

The First Lesson:
Opening to the Song of the Track

The Ojibwe people here in the Northcountry call March the Snowshoe-Breaking Moon. When the sun begins to climb higher in the sky and the first warm winds from the south lick the aging snow, its surface will soften during the day and refreeze at night. This usually forms a crust that can be walked upon—well, mostly walked upon. When you break through, the crust can give your shins a rap that smarts enough to make you wish you were wearing snowshoes. However, if you do, the crust shows no more mercy to snowshoes than it does to shins.

Julia, a student who has returned to the yearlong program for a second year, came up with another name—the Foxy-Lady Moon. It happened late one afternoon when she stopped by and asked if I'd come and take a look at some tracks she had just come across about a half mile back in the woods. "Some tracks" was a joke—the site was so trampled it looked like there had been a three-ring circus. But that is typical of soft-spoken Julia, a master of understatement.

I'll tell you the story the song of those tracks told us, the way I remember it.

Fox comes from the west, picking her way through the Alder-choked marsh. At the small stream midway across, she stops to listen and sniff the air for any sign of life. Buried under three feet of snow, the stream might easily go unnoticed. However, her keen senses tell her the waters flow as vibrantly as ever, and that a world of life thrives down there in the warmth and shelter provided by the snow.

She'd typically go up or downstream to find places where the snow dome collapsed and check the exposed stream banks for food, but not this evening. Something else drives her that is stronger than hunger and more insistent than her usual curiosity.

As she crosses the stream, her heart must be pounding. A headiness takes over that shows in the brief shortening of her stride and slight widening of her typically narrow straddle. Every fiber of her being tells her she is getting close. Her stride lengthens and her determination shows by her paws imprinting the snow more deeply than normal. It looks like no pain or distraction would deter her now as she brushes against twigs and squeezes ungracefully through tight places that she would otherwise have avoided.

Just before she breaks over the short, steep rise from the marsh, she pauses, lowers her head, and instinctively samples the previous day's damp melt air that lays heavy atop the snow. A near-full moon hangs brightly over the meadow just ahead, which makes it feel both safe and dangerous: she can easily see, which means she could also be easily seen. In the green season, the dense Sweet Fern shrubs would afford good cover for her rich orange coat, and at the same time she could easily peer through them. However, now no more than a few bare branch tips poke above the snow.

"Perfect," she thinks, "this is better than last year when the snow wasn't quite as deep." An ancestral voice within tells her she needs a broad, level surface in the middle of the meadow where her feet won't get tangled and she can see for a distance in all directions. It is essential that for this, one of the greatest events of her life, she feel safe and unencumbered. She knows she must be able to detect any incursion into the meadow and have much more time than usual to react.

Her tracks clearly speak her confidence; she has done this before, and she is merely shadowing herself. Flashes of sweet memories might be why she tenses with excitement and digs her toes into the crust while circling into the center of the clearing. The anticipation makes her light-headed, which is shown in her erratic track pattern. It has none of its normal catlike consistency. To make sure she is the only vixen around, she circles the windward side of the clearing with nose pointed upwind and alternately sniffs the scent-pregnant air and the snow. At this late stage of her estrus, she doesn't care if other males come and go—or not go—as practically anyone will do. This is all about her; to perpetuate her lineage, she *has* to become impregnated. It'll happen if she can only satisfy one need, one burning desire—the craving between her haunches. It is pure instinct at play, and this drama has only one voice: passion.

Satisfied that she has no competition, she feels the urge to pee. It is so strong that she dribbles before she settles into her usual squatting position, which is clearly shown in the sprinkling over the snow a step before her actual pee spot. The imprint of her rocking on her heels while she squats hints that her eyes are half-closed in near-orgasmic bliss.

Like water sucked up by a sponge, the faint skunky scent of her urine spreads through the meadow's blanket of heavy air.

Meanwhile, crossing the old logging road that runs through the Red Pine grove on the far side of the meadow is a male Fox. Last year he came upon this area, which he found rich with wetlands and mixed forest. He was from the farm country to the south, where there are more of his kind and thus less potential to find unoccupied territory and a mate. The trapper who kept this area free of "varmints" had just moved away, so this male happened upon the right place at the right time.

So did his mate, the vixen of this story. They had cubs last spring—probably his first mating—and then in the autumn she had mysteriously disappeared. She likely wandered off with her pups, who were looking for territories of their own. Her absence wouldn't have been terribly disturbing to him, as a Fox pair typically maintains only casual and sporadic contact during the non-breeding season. It is the way of the Fox for males to wander when there are no longer mates with pups to tend.

However, right now it is different. For the second time in his life his testicles are swollen, and he is driven by an urge that keeps him moving, keeps him keenly attuned to any sign of others of his kind. Where simple avoidance of other males was his modus operandi during the rest of the year, he now bristles at the faintest sign of one nearby. The casual interest he had extended to vixens he came across has now revved up to a blaring, single-minded obsession.

As he steps intently across the road, he picks up a light skunky scent carried on air that flows like a sheet of water down the slope from the meadow. The fact that he is barely able to detect the scent deters him not in the least. Hormones take over: his nose turns upwind and tows his body along to the crest of the hill where the Pines open into the meadow.

Forsaking his usual stealthy skirt around the clearing to get a feel for what is going on, he just steps directly out into the open. He doesn't even bother with a brief pause. To him this is neither a bold nor a foolish move, because the thought of having a choice never occurs to him. On top of that, he is oblivious to the fact that he is boldly exposing himself in the bright moonlight. It's simply the only thing to do.

The scent is now so overpowering that it has a reverse effect and shakes his rational processes out of their slumber. Somewhat back to his senses, he quits his beeline approach and his trail takes on its more typical variations. After all, his kind didn't survive this long by blindly following alluring smells.

Crouching a bit, he takes longer-than-average steps for the slow speed at which he is moving and cuts diagonally across the meadow to a little island grove of stunted Jack Pines near the other side. Not wanting to trust his hearing alone to scout the deep shadows under the Trees, and without the faint breeze in his favor to bring the secrets of the shadows to him, he keeps a healthy distance from the grove as he circles around to its backside. It seems as though he wants to take a good look at what is on the other side. Before he'll allow himself to respond to primal urges, he needs clearance from one or more of his keen senses.

She feels his presence before he rounds the grove, and the shock registers as a small ridge of snow ringing her rear prints. Considering the setting, she probably couldn't have detected him by sight, sound, or smell. She heard the song of his track.

Instinctively she squats to pee, but only a tiny dribble, as that is all her nervousness will allow. If there was any doubt in his mind about her readiness or his timing, witnessing that event—as clear a sign as a woman's inviting smile—is more than enough to convince him.

Immediately upon seeing him, she scoots over to a bush about a half-dozen paces upwind, appearing to be as shy as a sheltered farm girl on her first date. But it is only an illusion, as shyness is hardly her reason. This is just another step in the grand event she is orchestrating.

Rather than slinking behind the bush, which she might do if she were actually shy, she stops in front of it and turns her head to watch him approach. He zigzags toward her with nose to the snow as though he is sniffing for morsels. And that he is. Ignoring her completely, he goes straight for the dribble spot and with rapid sniffs intoxicates himself with its pheromonal essence.

Perfect—it is working, she thinks, and she gives a slight quiver.

He pushes his nose into the perfumed snow. The intensity and deliberateness of his stance and nose prints say in no uncertain terms that he has found the meaning of life. It makes sense that coming to such a profound awareness would be olfactory-based, as smell is a Fox's strongest and most trusted sense.

Without lifting his head, he goes straight to the larger urine print she made when she first came out on the meadow. While he sniffs, she comes forward tentatively, mistrustful. She needs her primary sense—smell—to confirm what her other senses are telling her about his gender and identity (and in this moment of passion, gender probably rates higher than identity). Part of her edginess is because Foxes are smell-oriented, and she is upwind of him, so she can't pick up his scent.

At the same time, she needs to know his interest in her is sexual in order to feel relaxed with him, and smell will tell her. Even though previously mated, a pair of Foxes typically goes through a lengthy mating ritual to reestablish their comfort level with each other. This just-reunited pair doesn't have time for such luxury if they want to have pups. She is probably at the very end of her fertile period, which typically lasts only a couple of days.

Obviously more interested in her urine than her, he ignores her until she comes up right beside him. With no more introduction than that, she turns her backside to him and raises her tail, which quivers and leans off to one side.

He is more than consumed with the passion of the moment—he is on a narcotic high, nearly oblivious to his surroundings. She likewise is in her bliss, yet she is the one responsible for their safety. She positions herself so they will copulate facing the closest tree line, because it is the nearest likely source of a potential threat, and it is the nearest refuge.

To support his weight and keep them both stabilized, she spreads her rear feet and digs in her claws. He mounts her deliberately and rises up on his toes to arch over her. She repositions herself several times in attempts to keep her balance, each time tamping her feet hard into the snow. Intercourse is over quickly, which the lack of definition in their imprints reveals.

After they uncouple, he licks himself and goes over to smell her vulva. She fidgets as though she is annoyed and takes a quick skip forward. He reapproaches her, cautiously this time. She stands still—grinning with ears laid back—signs of fatigue and resignation. And yet she feels a slight tingling in her groin. She cocks her tail off to one side, which arouses him just enough to make a half-hearted attempt at remounting her. This time her rear feet don't impress very deeply, nor do they dance around like the first time.

They came to the meadow as two and they leave as one, traveling at a diagonal to the wind in order to catch scent. Keeping about three body lengths apart, they give themselves a wide perimeter for awareness and safety—their first opportunity to practice the cooperation they'll need to raise their kits.

Like the eternal rhythm of waves washing up on the beach and returning to the sea, the pair comes together, parts, and comes together again. The vixen will stay with the kits continually when they are first born, and dad will do all the hunting. When the kits are about one moon old, she'll be able to leave them to go out and hunt down her own. At this time, dad will wander off to resume his solo life until he is again taken under the spell of vixen magic. After seven or so moons with mom, the kits will have learned the ways of the hunt, and she will send them off on their own, getting snarly if she has to. Like her mate, she will be responding to the pull of the tide by returning to the solo hunt to prepare herself and her territory for the time when he returns to wash up upon her.

2
Romancing the Frog

The Next Step: Hearing the Rhythms of the Natural World

Here in the Northcountry dwells an amazing creature called the Wood Frog. It is a fitting name, and Frogs in general have lucked out with good ones, like Tree Frog, Leopard Frog, Green Frog, Spring Peeper. Not so for Toads. We have Fowler's Toad and Woodhouse's Toad—who the heck are Fowler and Woodhouse anyway? And then there's Amargosa Toad—how many people know that Amargosa is an old name for California's Death Valley? A clear case of discrimination, I'd say.

Back to the Wood Frog. This little guy is under three inches long, comes in a variety of colors, and has a black mask. He likes large, uninterrupted tracts of moist forest, and he likes a lush undergrowth of Ferns and other greenery to keep the humidity high near the ground, which is where he lives. Like most other woodland Frogs and Salamanders, he likes to breed in ephemeral ponds. They dot the woodlands after the snowmelt and are usually gone before midsummer. Amphibians favor these ponds over open water because they harbor no Fish—voracious eaters of Tadpoles. Also, these little pools are often the only close-by water.

Because the ponds are short-lived, Wood Frogs can't afford to waste any time waiting for warm weather to breed—or even waste time courting, for that matter. If Pollywog and Tadpole are going to grow up to become Frogs before the ponds disappear, Mom and Dad have to get right down to business. They have a window of only a few days or they miss their opportunity for the year. So it is pretty much a case of, "Hi, my name's Herman, wanna have sex?"

One sunny afternoon in early spring—and I want to stress the *early* here—my mate Lety and I were sitting on the grassy bank of a little pond that was still mostly frozen over. The north side of the pond where we were sitting had managed to absorb enough sunshine to let go of a thin strip of its icy shield along the shoreline. There, right before us in the shallows, a slight movement on the surface caught our attention. The top of a head emerged and remained stone-still, glistening in the sunshine.

Moments later we heard what sounded like the quack of a sick Duck. It wasn't nearly loud enough to be a Duck, and besides, it was coming from the Frog. So it probably wasn't a Duck. After a couple more evenly spaced quacks, we heard another quack, only more subdued, coming about an arm's reach away from the Frog. He moved ever-so-slightly in the direction of the new quack. Again he would quack, and again it was answered by the softer quack, and again he edged closer, this time by a couple of inches. This quack-respond-move routine continued until he was on top of her.

Now that's bare-bones reproduction. No endless nights of trilling or croaking or booming to attract a mate, like other Frogs have to do. No romantic nights on a moonlit lily pad. Not even any foreplay or afterglow. "Highly overrated," says the Wood Frog. Just unthaw (he was frozen solid all winter), crawl out from under the leaf litter, shake your legs to work the sugary antifreeze out of your veins, head right down to the nearest pond, paddle over to the closest female, and voila, your procreative duty—and procreative pleasure—are over. *And* there's still time left in the afternoon.

It's all about business. They shook themselves out of their long slumber for one reason: to bask for a moment in the sun-drenched icy water, and then lose themselves to the rapture of communion.

Were they thinking about the continuation of the species? Were they even concerned about being watched? My hunch is that all they knew or cared about was the now. I doubt there were any dreams about how their children might turn out, or about their own places in history. They were there at that one moment in time to serve the continuum of life. Sure, they were leaving their track in the pond, and in doing so they were leaving their track in time. But that only mattered to Lety and me. The Frogs had more important things to tend to.

The song of their track fascinated me. Part of that song was in my memory, as I've watched the Wood Frog perform this ritual in seasons past. And I knew somehow, deep down, that when a new track is laid it is not really a new track, only another singing of the song that has been sung for millennia. No wonder I drift into a timeless state when I'm tracking. I am track maker, tracker, and tracked all at the same time, just like the Frog, and just like his Ancestors and mine since the dawns of our species. To say that tracking isn't in our blood would be like claiming the sun doesn't shine.

I looked at the track he left in the floating plant matter and I was humbled by the thought of how many times his father, his father's father, and on back have made the same track, on this very pond in the same place, on the same day of the season, and for that same reason. I realized that the songs of the old tracks were still here, still echoing in the pond, the surrounding forest, and beyond. It was as though they were all singing together in one giant chorus, with each other's songs separated only by the turns of the seasons.

I wonder how many times the Wood Frog mating ritual on this pond has been observed. This too is an aspect of the song of the track. This time around, I am part of the circle of the track. Perhaps another time it was another man, and another time a hungry Mink, and so on. A Wood Frog breaking the water's surface and belting out a poor imitation of a Duck quack is bound to draw attention. But they don't seem to care. There is risk in life, and there is risk in love. And this is true especially for them, because the smaller the window to love, the greater the risk. I wasn't hungry that day or I could have participated in the song in a different way, as the Mink did with this Frog's great-great-grandfather a few years ago.

The more time a Frog has for love, the more conscious he seems to be of the song of his track. I recall one night when I was sitting with a few students around a small fire on a little knoll overlooking a bogland pond. It was early in the Budding-Leaves Moon, before Leeks and Spring Beauties turned the brown leaf mat to a plush carpet of green and pink. Legions of tiny brown Frogs known as Spring Peepers had convened down at the water. From dusk till dawn at this time of year, they raise a high-pitched shrill to collectively express their amorous longings. It fills the farther places with a din of such volume and brain-piercing pitch that it could drive a person to tears or worse.

As we sat there listening and chewing on some tough jerky, we noticed pockets of silence in the chorus. Someone asked why.

"Open to what the voices are telling you," I suggested. "Listen without thinking or concentrating on the sound."

"There's an eerie pattern," said Sarah, a young, outgoing woman from Boston on her first wilderness adventure. "That quiet pocket over there is moving through the Frogs, like a slow wave over the ocean. What is causing it?"

Around the fire the others offered ideas: a chilly breeze, a chain reaction, a breather in the communal mating frenzy.

"Let's listen some more," someone said.

Tom, a self-reliant outdoorsman from Texas, noticed that rather than progressing smoothly and steadily like a water wave, this one varied in speed and intensity. "It acts like it is up to some purpose," he said, "like it has a life of its own."

I watched chills crawl up a couple of spines.

"It's a Fox stalking the edge of the bog," I said out of the blue. Immediately I had everyone's rapt attention.

"How do you know?"

"Yeah, what do you mean?"

"If I thought I knew for sure, I'd be a poor tracker. When I think I have something figured out, I quit asking questions, and that shuts me off from the song of the track. The more I'm proven wrong, the closer I get to what is actually going on. My Fox statement was only my truth of the moment. I'm listening as we speak, and if the circle of silence grows bigger, it could be a Raccoon—they create quite a disturbance when foraging. The more the circle moves in fits and spurts, the more it resembles the erratic energy of a Mink. Or maybe the Frogs are just playing with our minds."

They sat silent, absorbing this. There were hints of skepticism on a couple of their faces.

"In grade school," I continued after a bit, "do you remember how you'd quit talking and goofing around when the teacher came by, only to start right up again when she turned her back? This is what the Frogs are doing. It is called a blackout, and sound, visibility, and movement blackouts are common occurrences in the woods and the city. Songbirds will quiet and hunker down when a hunting Hawk moves through their area. Isn't it amazing how incredibly varied creatures are, and at the same time how incredibly similar we can be to each other?"

"I'm missing something here," declared Dan, a studious social studies major from Southern California. "Let's say it *is* a Fox: how the heck is any Frog gonna hear him in the middle of this horde of screaming banshees?" Dan grimaced and rubbed his temple to make sure we knew he was getting irritated by the din.

"It's all about relationship," I replied. "As you may know, relationships—healthy ones, that is—are dynamic and ever-changing. And they're many faceted. Anything I, as an outsider, might observe can only be a piece or two of the puzzle. If I claimed to know anything more, I'd either be a fool or a liar. I'll give you a little peek into the relationship here as I know it, but that is all. The rest is for you to discover, for the same reasons you'd probably like to develop your own relationship with your lover rather than have someone else do it for you."

Dan nodded in agreement, so I continued.

"Think of the Fox and his mate and newborn pups as one life, and the Frogs as another life. They need each other in order to be Frog and Fox; they keep each other healthy. When Fox comes near—even before he can be heard or seen—Frog picks up on the song of his track and grows quiet. This makes her invisible, even when she is in Fox's reach. At the same time, Frog keeps Fox surrounded with a chorus of song so loud and harmonious that he can't key in on the source of any one voice."

"So the Frogs together are functioning as one life, I can see that. It's not quite as clear with the Fox family."

"Perhaps if you were to become the Fox, you could get in touch with his motivations for being out here tonight."

"Do you know what they are?"

"I know nothing; all I do is listen to Fox tell his story. And it's not mine to tell; you'll have to hear it for yourself."*

Dan laughed, shook his head, and let his eyes follow mine out into the story of the night.

---

*For more on the teaching approach in these stories, see appendix 1.

3
Bear Stump

The Tracker Starts to Read the Stories of the Forest Floor

A light frost glazes the bog, which is not uncommon here in the Northcountry even though it is the Blackberry Moon, the warmest time of year. Right now the first rays of sunlight are touching the ridge I just followed to get around the bog that lies at the foot of our wilderness camp. There Greg should be waiting for me. He hired me to guide him on a one-day nature study and tracking intensive, and as always, I'm raring to go. Ever since my inaugural gigs with my little brother fifty-five years ago, I've thoroughly enjoyed sharing the mysteries and beauties of this land I call home.

I have Greg for only a day, so I'm going to do what I can to make it a rich and inspiring experience. He is an engineer with a growing family and not a lot of opportunity to get out on his own. As a child he used to come up to his grandfather's cottage on a nearby lake, so today holds special meaning for him.

There will be a lot coming at us and not a lot of time for follow-through, so we will want to solve riddles as they come up. This will be an out-of-the-ordinary day for me; I won't be playing the subtle guiding role I usually do with long-term students. On top of that, I'm going to be an interpreter and tour guide, as much of the wilderness and its voices are going to be new for Greg.

At the same time, I cautioned Greg that if he came with the expectation that I was going to play naturalist, he'd be disappointed. My way, I explained, is to help people discover things rather than point-and-identify. I told him straight out that I am not a tracker or a hunter, or an ecologist, or anything else, for that matter. I practice all of these skills, just like most natives. However, like them I don't specialize in anything. A specialist can do very well when he hits his groove, but he could falter when variables come up that lie outside his realm.

Because a native is a generalist, he has a broad base of knowledge and experience that makes him better at specialized tasks. For example, when I'm tracking a Bear, I draw upon my knowledge of the Bear clan.

I know they like to travel the edge of a wetland because that is where they find their food. I know what they eat because I watch them forage, I read the sign of their foraging, and I study their scat. I have a personal relationship with their food because I eat it myself. Knowing the fruiting time of their food plants, I know where Bear will be and when. Being able to forecast the weather, I have some sense of when Bear will be moving and when she'll be laying low. The signal animals, Raven, Jay, Squirrel, and Deer, are my scouts for Bear. In knowing their language and habits, I learn more about Bear's. On a hot afternoon, a swarm of Deerflies can show me where Bear is napping before I get close enough to disturb her. Even my understanding of glaciation and soil types helps me know Bear's moods and motivations. Areas of rich clay soil formed by glacial outwash attract a hungry Bear to the early spring greens they support, while a mom with cubs spends a lot of time later in the season in the berry patches that flourish on sterile, gravelly glacial till.

As I skirt the bog, I marvel at the emerald green of Calla Lilies growing amongst the Blue Flag Iris. Two Ravens fly up and give a melodious *ga-rawek* to herald my arrival. Old Red Squirrel, who knows me well, gives me a chatter of recognition that barely interrupts her pinecone shredding. It appears I didn't surprise her—which I like to do—as my shadow walking was picked up by others besides Raven before I got this far. Were I with another person, she'd have raced up a nearby Tree and acted as though her territory was under siege. Scolding and taunting us, she'd send out an alarm cry that would be picked up by one Squirrel after another and reverberate through this part of the forest.

I ask Squirrel for a couple of the remaining blueberries at the base of her Tree and she obliges. Ignoring her presence and approaching indirectly so as not to threaten her, I act as though I'm intending to just pass by. My hand unconsciously brushes the bush and the berries make their way to my mouth, unseen by anyone who might be watching. If I showed any focus on berry picking, Squirrel might feel threatened and raise a ruckus because I was hanging around in her territory.

About twenty paces ahead, two Deer cross before me: a yearling and her barren mother. Were she still fertile, she'd now be with a new fawn and would have already driven her adolescent child away. For some reason the two are overdue for their usual early morning trek across the bog to browse on late-season Violets under the far Maples. The Violet leaves, though still succulent, are looking ragged from the many Bugs who feast on them. After their snack, mother and daughter will probably descend the north side of the hill and nap in a thick grove of Balsam Fir to escape midday's biting Flies.

I know this about the two Deer not just because I know them, but because I know their clan. The sign of their patterns chimes out like Blue Jay's alarm call. There's the trail they use to cross the bog that is so worn down that water stands in it year-round, and nearby bushes and low Tree branches shaped by the Deer's browsing, and out-of-place plants flourishing in the otherwise-sterile peat where the Deer regularly deposit their nourishing scat.

Had I not been here for several turns of the seasons, I'd still know their ways. If I were not to return for several turns of the seasons, I'd know the ways of their grandchildren. The cycle of their lives and the pattern of their movements remains the same. The clan goes on living, with individuals seeming to endure for generation after generation by being given new bodies. The daughter literally walks in the footsteps of her mother, as will the daughter's daughter, and so on through the life of the clan.

When these Deer I have come to know no longer wander the forest with me, I sometimes visit their bones in the quiet place where they go to die. Though a body withers with each death, the clan freshens with each birth and walks the timeless continuum across this bog and back again.

A social animal has her intelligence centered not in her individual brain but in the clan consciousness. Each member of the clan contributes to this collective intelligence, which allows the whole clan to benefit from each member's insight and experiential memory. We see this exhibited in the think tanks and brain-storming sessions of our species. We see it in a school of Fish or a flock of Birds moving together in choreographed synchronicity. This is known in native communities as clan knowledge.

The two Deer who just passed before me exhibit it in their daily routine. In the white season, when all other sources of drinking water are frozen up, the Old Ones remember that water appears on the lake above the ice, and that it can be reached by pawing through the snow. The weight of the snow upon the ice forces water up between cracks. Pooling on the surface, the water can get a hand's width deep. The Elders teach this to the young, who in turn pass it on, and thus it continues on as clan knowledge.

The Elders help span the time between events that occur sporadically, such as droughts, fires, or exceptionally heavy snows. They hold this as part of the clan memory, so that it can benefit the coming generations. For example, if it were twelve years between droughts, those less than twelve years of age, even though they might be grandparents several times over, would not know the trails that could lead to water. However, those more than twelve years of age would.

Clan knowledge is cumulative knowledge, with each generation contributing to it. An individual might discover something new, which contributes to the good of the whole. If he is able to pass this knowledge onto others, it becomes part of the clan's body of knowledge. In this way the clan grows in lifeway skills, as well as in the wisdom drawn from them.

Much like our human oral histories, clan knowledge has a life of its own. In order to persist and remain vibrant, it needs the generational dimension of the clan, which includes Elders, adults, and young. Imagine the havoc that would be wreaked if the rhythm of the continually unfolding seasons was interrupted by a season that failed to appear. If ever the Deer clan were to miss a season, such as all the young—

the learners—dying of disease, or those in their middle years—the teachers—being over-hunted, or the Elders—the wisdom-keepers—all dying in a harsh winter, the clan knowledge might not get passed on.

This has happened with much of our human clan knowledge. When our Eurasian, African, and American hunter-gatherer Ancestors were conquered, our Elders were silenced, our parents were enslaved, and our young were indoctrinated into foreign ways. The continuum was broken and our knowledge quickly died. It can happen within the space of a single generation.

The reverse is also true: clan knowledge can spread from clan to clan. When a Deer migrates to another area, she might share some of what she knows with her new kin. If it works and is accepted, it will become part of their clan knowledge.

Animals new to an area uninhabited by their kind face a unique challenge. Imagine Deer colonizing a new area or being reintroduced to an area where they were exterminated. They wouldn't have the clan knowledge suited to their new area, such as where to find water in times of drought or after freeze-up. They wouldn't have an established migration pattern, along with many other knowings that made life good for them when they walked in the shadows of others. I think this lack of clan knowledge is one of the primary reasons many animal introductions fail.

Some of us who hunt choose the biggest, most magnificent animals of the clan—the trophies. In their prime, they're the leaders of their clans, the bearers of bright and healthy young, and the teachers of the clan knowledge. This is what our trophy costs the clan. Wolf, on the other hand, usually takes either the old, who have already passed on the clan knowledge and no longer have the strength to lead, or the young, who haven't yet received the clan knowledge and are usually born in numbers greater than the environment can support.

Lack of awareness of the ways of the Deer clan that results in trophy hunting reverberates through our human clans as well. Our Elders possess clan knowledge that is being lost because we mistakenly believe it doesn't fit in our world, which we see as so different from the one they grew up in. The upshot is that we're not learning what was passed on from our Ancestors, and we're not faring well without it.

The two Deer pass before me and the old doe looks my way. This is her responsibility, and the young one trusts in it. I move with her way of moving and I wear the skin of her kind, yet she is nervous because I stand upright. I avert my gaze and get down on all fours to browse. But I'm too late—she woofs, flicks her raised tail, and bolts smartly off into the woods with yearling in tow. Spooked as she is, she won't cross the open bog. Along with exposing herself, the spongy turf would slow down her escape.

After I pass, she'll likely circle back around, as is the way of her kind, and she'll nervously approach the same spot. She'll be keen for any sign that I might still be around, and once satisfied that I'm gone, she'll take her offspring across the bog and they'll continue their day as usual.

The first thing Greg tells me when I walk into camp is that he already had breakfast and he is set to head out, which gives me a quick smile. Such enthusiasm right off tells me he is bound to have big fun, and I'm already in a state of bliss, so I couldn't imagine a better way to start the day.

We read the clouds, the wind, and the humidity, and we compare them with yesterday's to forecast what we have in store for today. It looks like no rain or falling temperature, so we leave extra clothing behind. As always, I strive to travel as light as possible. I suggest to Greg that he drink the contents of his canteen and leave it behind. The most efficient way to carry water is in your belly. And when you're active, it's good to drink before you get thirsty, because when you feel thirst, you're already getting dehydrated.

We head south under a dense canopy of Hemlock and Cedar. It breaks at the edge of a bog, where we come upon several knee-high ridges around two to four paces in length. The ground around them is sterile, supporting only Reindeer Moss, a few Sedges, and the occasional scrubby berry bush.

"This land looks bulldozed," says Greg.

"Are you sure?" I ask. "Look around and notice what is unique about this area."

He points out that the ridges are all similar to each other and lay parallel to each other right at the edge of the bog. There's water standing below the ridges, which seems odd to him because the bog appears dry further out.

"The breeze is light," I say. "Can you still pick up the wind direction?"

"Yeah," he answers. "I can feel it on my face, coming from across the bog."

"That's typical; what direction is it coming from?"

"Well, the sun is rising behind me and a little to the left, so I should be facing southwest."

"Okay, I think you now have all you need to solve the first mystery of the day. There's an exercise called *becoming* that will help you put the pieces together. Don't think about your question and don't try to figure it out. Just let yourself melt into the wind and the dirt and the water. Don't focus on anything. Take it all in and let yourself feel, let yourself envision. When everything comes together, it is like the jumbled pieces of a puzzle assembling before you, and there you have your picture. Do you want to try it?"

The first time doesn't work—he starts to think. With no prompting, he does it again.

I barely get to sit down and get comfortable, when without saying anything, he walks about ten paces to the north, where there is a downed Spruce Tree with its upturned, weathered roots sticking up about as high as we stand. He looks back at me with a knowing smile.

"Tell me the story," I say.

"Hmm. I'm not sure how accurate this is, but here's what came clear as soon as I saw the picture. This ground is just sterile sand and rock, and there's plenty of water near the surface, so there's no reason for the Tree to grow deep roots. Instead, they spread out to find nutrients. It shows in these roots—look how they're splayed out flat as a pancake. When I was becoming, as you call it, I saw this root tangle rotting down and forming another ridge. The prevailing wind tips the Trees all in the same direction, which is why all the ridges run parallel to each other. The upturned roots at the edge of the bog leave depressions, which fill up with water. It's funny, Tamarack, how quickly everything had a reason as soon as I could put all the pieces together."

We're off to a good start. Rather than a focused find-the-answer approach, Greg's becoming put him in an open, questioning place. A good question is worth a handful of answers. An answer is a piece of knowledge—a dead end, where a question creates a thirst for knowledge—an open end. So often an answer comes from either a good memory or good sources, where a well thought-out question shows intelligence and perceptiveness. Such a question points to its answer, and the reward is all that is discovered along the way, even if no answer is found. In fact, more often than answers, I find other questions that take me down alluring paths I couldn't have imagined when I started.

"We now have a choice," I tell Greg. "We can either cross the bog or follow its edge."

Greg opts for the edge, which promises to be an adventure, as it'll take us across a narrow isthmus between two bogs that serves as a major animal corridor.

Sure enough, Greg's attention is soon caught by an upturned piece of rotted log, which he stoops over to examine.

"It's easier to see the forest when one's nose isn't tight up against a Tree," I suggest. He steps back and notices the stump from which the log came, which is partly torn apart as well. Around the log winds a sandy trail, which is peppered with tracks.

"Why such a busy trail right here?" Greg asks.

I'm glad we're starting with a question. However, this riddle promises to be tougher than the first, and we're no doubt going to be given many more throughout this day. Still, I continue to smile, as challenge is really opportunity in disguise, and it is questioning that removes the mask. Greg came here because of his hunger to know the native within him, and I'm honored to serve his awakening in what way I can. We are native when we let ourselves be native, and today Greg takes another step.

We backtrack to the beginning of the isthmus to see how such a well-worn trail came to be.

"The trail disappears!" exclaims Greg. "Oh, I get it! This skinny neck of land is like a bridge across a river, where everybody gets funneled together to cross. I would guess the trail's going to disappear at the other end too, because everyone's going to go their own way."

Greg is curious as to who is crossing on the bridge, so we return to the trail section around the log. Three tracks are more distinct than the others: those of an adult Deer and two very small fawns.

"Why are these tracks so clear when the others are just pockmarks in the sand?" asks Greg.

"Try becoming again," I suggest. "Walk the trail and make the tracks. But don't think about making tracks—just *be* the experience."

He closes his eyes, takes several slow, deep breaths, and tells me this story: "It's spring and the ground is mushy from the sun thawing it. We're out here on the isthmus because it's open and the sun feels warm, and it's a good place to find some early greens. The trail is slick from the frost, so we walk along its edge. That's why the tracks haven't gotten trampled."

Greg opens his eyes and breaks into another knowing smile as his gaze returns to the disturbed stump and log. "I wonder. . . ." he says. "Maybe the Deer stumbled over the log trying to stay off the slick trail."

I can tell he needs more information before doing any envisioning, so I ask him to observe all he can about the scene without disturbing anything.

"Well, these chunks of rotten log are just overturned," he says with a thoughtful look, "but the stump looks like it was torn apart."

"Do you see any relationship between the two?"

"I just assumed they were done by the same animal at the same time," he replies. His pursed lips and drooping stare show a hint of frustration. "Ah! I just did it, didn't I? I gave an answer instead of asking a question. If I learn one thing today, it's going to be *don't speak; listen.*"

"Hey, don't beat yourself up," I say reassuringly. "We learn something well only by making the mistakes that teach us what does and doesn't work."

"Point taken," says Greg. "Now, how can I pin a time on each of these disturbances?" he asks. "I have no idea where to start."

This guy is genuinely stumped—too stumped for becoming to work. If you're going to put a puzzle together, you need the pieces. I suggest that he pick out each clue he can identify, not to examine but to see how it fits in relation to its environment. "It's all about relationship," I tell him. "If a woman brings me a clue like, oh, say an eggshell, I'll ask her to take me to where she found it, we'll replace it just like it was, and then we'll stand back and listen to the story. It's not the eggshell telling the story, but the eggshell's relationship with its Hoop of Life."

Greg gets to work. "Look here," he says. "The Lichens are dead and rotting under this chunk from the log, but they're just bent over under the stump pieces. The log was disturbed at an earlier time."

The fresher stump disturbance draws his immediate attention. He attempts to envision the scene, but to no avail. So he becomes the animal and gets down on his hands and knees to crawl down the trail toward the stump.

While he is doing that, I try my own envisioning. When I am on the move and there is a lot going on, I'll envision. It is easy to slide in and out of, and I can adjust it quickly to fit the situation. Becoming is my choice when I want to gain an intimate feel for another's feelings and motivations.

My envisioning takes me back two moons, to the early green season when the Ant colony in the stump is a flurry of activity. I, a yearling Bear, come up the trail and sniff at the base of the stump, as is my kind's custom. I hear the rustling inside the stump and rake my right paw across the rotten wood. I do it again and again but only come up with a few goodies. Spoiled by easy meals in this lush season, I soon ramble on.

I know about the sound in the stump and what it means from my initiation into the clan knowledge by my mother. She has unique ways of doing things that were passed on from her mother. Other clans probably have some knowledge that is different from ours.

As Greg nears the stump, he realizes that he is coming from the wrong direction. The stump damage and debris are on the other side. Ambling up the trail from the other direction, he is clearly the Bear. I can smell him. I can see his head swaying back and forth to catch the scent on the breeze. Now I know he'll solve the mystery.

Greg raises his right hand, now a paw, and scrapes down the side of the stump. As he does, he sees the claw marks. It doesn't seem to surprise him; it is as though he expected them to be there. He looks back to where the scrapings would have landed, and sure enough, there they are. He goes through the clawing motion a couple more times, to gain a better feel for the cause and effect of his actions.

Greg asks about the size and age of the Bear. It is something that is beyond his present ability to determine on the spot. I suggest that he measure the width of the paw scratch by the width of his hand, so that tomorrow he'll be able to find the Bear's paw size on the aging charts in our library resources. Here I offer more than my usual background guidance to encourage his passion rather than have him hit a dead end so early in the day.

Additional information on Bear behavior could come from many sources. One option would be the naturalist approach, where we'd consult tracking books, field guides, and other resources. We'd be approaching Bear from the outside in, where if we became Bear, we'd be approaching her from the inside out. This would give us the opportunity to know her rather than only knowing about her. We'd be able to walk in her paws rather than just study her pawprint. The books can still be of value, for sure, but as complementary rather than primary sources of information.

Now Greg starts to worry me with this Bear thing he has gotten into. He lumbers on past the stump to a patch of Moss heated by the intense morning sun and plops over for a nap—just as the yearling Bear probably would have. Bears are masters of the nap. They consider getting in a dozen snoozes before bedtime to be a good day. Being a napper myself—though quite an amateur compared to Bears—I don't have to think twice to follow Greg's example.

Without Bear's black coat, which keeps her cool by absorbing sun's heat on its surface and radiating it off, we soon overheat. Greg leads us to the other side of the isthmus, which is shaded with a dense stand of stunted Spruce. However, the coolness comes at a cost for Greg. He charges head-on into the tangle, flailing his arms against branches and leaning back to protect his face from their rebound. He's not the cussing kind or I'd surely be hearing it by now.

I show him how to stoop low and lean forward to snake through and under the branches. It is easy going near the ground, where the branches have died off, and it's clear to see ahead. Our animal Relations know this, which is why they can so easily disappear into the brush. When someone attempts to navigate a tangle upright, they can get knocked off balance, get slapped in the face with branches, and end up with debris in their eyes.

Rather than reaching out to push branches away—they'll whip back at you, guaranteed—I suggest that Greg keep his hand right in front of his face. I tell him not to worry about branches hitting the person behind him—in this case, me—because trail etiquette states that it is the responsibility of the person behind to keep enough distance to avoid getting slapped. The person in front is then free to keep his attention focused on the trail. Some Boy Scouts would disagree. They think they should either warn the people behind them or hold the branch so they can grab it. Personally, I think these lads have taken the line in their oath to help other people at all times a little too far. It can detract from their experience, and it contradicts another line in their oath, which is to keep mentally awake. The person behind gets complacent when he trusts that the person ahead will watch out for him.

A few scratches later and we're on the other side of the isthmus. "I wonder," I say to Greg, "if there's something else that drew you over here besides the shade. The gatherer and the hunter are often guided by intuitive voices to what they seek. These voices are unexplainable—and sometimes not even recognized—yet they're there, and they work."

Greg looks around. "Could it be the stump?"

He is referring to the mostly rotted-away Cedar stump at the edge of the bog not far from us. This time he knows by the size and character of the disturbance around the stump that it was likely done by a Bear. He doesn't have to reenact it to see how it happened. Yet there is more here that I don't want him to miss. He reads in my facial expression that I'm looking at something beyond the obvious.

As though he's always been doing it, he stands back to gain perspective.

"The Bear followed the shoreline from the right," he surmises.

Greg is right on. Bears like to comb the rich band where two distinct habitats meet. Ecologists call this an *ecotone,* and I refer to it simply as *edge*. The lushness, known as *edge effect,* comes from the merging of plants and animals from both habitats, along with some species unique to the edge. This shoreline, which is a meeting of bog and upland, has thickets that are typical of edge. They provide sheltered nest sites, loafing areas, and travel routes for a variety of animals, like our Bear. When they venture out in the open to forage, they can duck into the thicket at any sign of danger.

The extensive nut and berry patches you often find along the border of edge are the result of animals—Birds especially—spreading seed through their scat. The seedlings thrive in the combination of sunshine, moisture, and nutrients that edge provides. This particular edge is rich in Blueberries, Raspberries, Willow, Lichens, and a variety of succulent plants, which attract a range of animals from Aphids to Moose.

"And it looks like he did the same thing he did to the last stump," Greg adds.

"Yes, and there's still more going on," I reply. "There's what I call an envelope, where there's one event, the envelope, which disguises another one tucked inside it. Look at the stump as the envelope. Look into it and follow it back to see what it holds."

This is a very old stump, the last remnant of the large Cedars that used to cover this isthmus. Greg and I can find only the vague impression of a long-ago fallen Tree, in the form of a slightly elevated mound reaching out into the bog. For the trained eye, it is easy to spot, as the drainage and fertility provided by the rotted wood supports upland plants such as Bunchberries and Sedges that wouldn't typically be found in the bog.

As for the stump itself, it can hardly be called wood anymore. It has been weakened by rot and fractured by frost to the point where it barely has enough structure to support itself. It's so broken up that it looks like the creation of a child stacking little odd-shaped blocks. There are no Ants in the stump and there haven't been for a while. It is too deteriorated to support them—water runs down into it like it is going through a sieve, and predators have just as easy access.

"Still, this doesn't make sense here," says Greg, referring to the pile of rotten wood that is too far away from the stump to have just sloughed off. "I don't see any obvious Bear sign, like the first stump, but still I don't know who else it could be."

Skunks will forage a rotted stump for little morsels, burrowing small holes for their sensitive snouts to poke into. And if there is a nest of succulent baby Mice to be had, a Badger will charge in with both of his front shovels going at once, fanning debris out on both sides. As with the last stump, the debris lays off to one side only, in a tight, orderly-looking pile.

"This one's got me," says Greg, sighing. "Defeated again." Then he quickly adds with a smile, "Just kidding! But I really don't know where to start."

This isn't like the other stump: the ground around this one is covered with short, thick grass, and there is no obvious trail.

"This Bear—if it was a Bear," says Greg, "could've come from any direction. And this wood—if you can call it that—I don't see it holding claw marks."

He is right. The "wood" is so leached out that it is as light as foam and crumbles to the touch.

"Remember the benefits of being a generalist," I advise. "You can take the perspective of a botanist or a weatherman or a fool rather than a tracker. And don't forget to embrace the Zen."

He has got this figured out—I see the gears turning as he searches for the clue in my suggestion. He scans the sky, the surroundings, and the plants growing around the stump. Then he steps back, takes a few slow, purposeful breaths, and comes crawling back to the site as Bear.

Meandering across the bog toward this little open spot in the wall of Spruce, he doesn't notice the stump until he nearly reaches it, and still he doesn't make a beeline for it. As Bears do, he sniffs and snortles his way over, just in case he comes across some tidbit along the way.

The approach doesn't feel right. He comes up the shoreline from one direction, and then the other, and finally his movements fall into sync. As he half-heartedly claws at the stump (there probably weren't any hidden treats anyway) he snaps out of being Bear.

"I think I have something," he says offhandedly to conceal his excitement. "That Blueberry Bush poking up through the rubble shouldn't be there if the Bear came through recently. He must've been here a while ago."

"Oh yeah?"

He gets down and studies the Blueberry Bush, being careful not to disturb anything around it. "Would it be okay if I move some of this debris from around the plant?" he asks.

I see no harm, I tell him, as long as he takes a mental picture and reconstructs the site as best he can.

To me this is an archeological site, and I wish to keep it preserved for posterity. This is my way with any natural sign that is not of my making, whether it be a track, a Bird's nest, a Beaver lodge, or a spiderweb. I am careful not to assume that my desires ought to take precedent over my nonhuman Relations. I strive to disturb them and their creations only out of need, and not merely for want. This is the Honor Way of the natural realm. This is how the Relations regard me, and how my heart guides me to regard them.

"This plant was here when the Bear came," Greg discovers. "It is rooted in the ground below the pile; it got buried."

"How long ago was that?" I ask.

"Whew, beats me."

"If you could live with Blueberry for a few turns of the seasons, she'd be glad to show you how she keeps track of time. We're only given today to be together, so I think it'll be okay to go with my experience with this one. If you'd like to reconstruct the scene, I'll describe how Blueberry does it. She is like most woody plants; she sends out new growth each year from a terminal bud. The bud scar usually persists, so the age of the branch can be found by counting the number of bud scars up its length. When the bud scar disappears with age, annual differences in the bark might still show, or there may be side branches that have sprouted from the scars."

"Four years ago," states Greg firmly. "The plant grew up through the rubble for four years since it was buried. I . . . ah, the Bear, was here four years ago."

"And?" I add teasingly.

"What? Isn't the riddle solved?"

Greg already knows not to take my every word seriously; however, he sometimes has trouble knowing when I'm serious and when I'm not. Nearly everybody does. I suggest that they take only about thirty percent of what I say seriously, and—here's the catch—it is up to them to figure out which thirty percent. I do this sometimes for fun—*humor in all things* is my motto. However, most of the time it is to keep those I guide sharp. It helps them to be inquisitive and think, and it trains them not to assume anything, not to take anything for granted.

Now this is just between you and me—I don't want my students to find out—but I'll often add another layer of confusion by keeping my teasers subtle. This makes it harder for them to tell whether or not I'm serious, which keeps them all the more in questioning mode.

This can be tricky business. I give serious consideration to each time I use the technique, as inappropriate jesting can be disrespectful, and it can backfire. I need to be sure I'm not abusing my role as a guide and doing more harm than good. We naturally want to know—and sometimes we need to know—whether or not we can take what somebody says at face value. Honor and respect are my guidelines, and I play the role of jester only in the context of a learning-based relationship.

Perhaps my biggest challenge is to keep the focus on the subject matter rather than me. Doubting and questioning me is just as important. "Don't take anything I say as gospel," I tell them. At the same time, focusing on the teacher—even if doubting or questioning him— can become a trap. The trackers' mantra is not "Question," but "Be as a question." Be is a state of being, which in the natural realm manifests as a questioning approach to life. This translates as curiosity and a desire to learn and grow. If instead we focus on the questioning aspect, some of us will have trouble separating the person from the topic. We can end up developing a habit of doubting and mistrusting people and situations, and even ourselves.

If I were to come out directly with something like, "Hey, wake up! You're lost in yourself; you're caught up in your own reality," I'd be playing a common teacher or parent role. It could trigger defensiveness or anger. Or it might create a situation where the student becomes dependent upon the teacher's input. Well-executed jesting, on the other hand, can trigger an emotional response that encourages exploration, enlivens the senses, and quickens the intellect.

My simple "And?" is all it takes to keep Greg questioning. It is questions that awaken our tracking abilities and guide us on the path to attunement. If Greg quit questioning, he'd probably have treated the track as a message to read and been done with it. He may have missed a step in the transition from reading track and sign to hearing the song of the track. Imagine a Wolf stopping to study her own pawprint. When we become the Wolf, looking at the track could be just as redundant.

Then why do we do it? When we have lost the ability to be as a question, we try to compensate by becoming analytical, i.e., looking for answers. Our culture teaches us that if we take whatever part of an organism we might have and study it enough, we can get to know the whole organism. Perhaps. I suppose someone could analyze my hand thoroughly enough to find out a lot about me; however, what's the sense of doing so if I am right there connected to my hand? In the same way, a track is connected to the trackmaker.

Another way to grasp this is to think of the track as a chapter in the middle of an endless story. If I'm sharp, I can study that chapter for clues as to what happened in previous chapters. And with a bit of luck I might even be able to predict what is going to happen in the upcoming chapters. However, I won't be in the now. I'll be in the past when trying to figure out what already happened, and I'll be future projecting when attempting to predict what is going to occur. In the now, past, present, and future take place simultaneously.

So many of us deprive ourselves of this communion because we suffer from what I call the Sherlock Holmes Syndrome. Like ol' Sherlock, we take a clue and deduct everything we can from it, because it is all we've got. The field of archaeology is built on this approach. When the archaeologist has no feel for how a hunter-gatherer people lived, he tries to deduct it from an artifact. His analytical mind keeps his intuitive awareness, his ancestral memories, and his sensory acuity at a distance. Were it not so, perhaps through our shared human experience as hunter-gatherers he could remember when he lived there with them.

And the same goes for many of us and the tracks we find. Sometimes we treat them like an archaeologist does an ancient vase or oil lamp: we photograph, sketch, measure, compare, and plaster cast them. What are we doing? Perhaps we are trying to use mainly our heads to become the trackmaker when the ability lies in our heart-of-hearts.

Greg listens, and then he looks again—not at the sign that exists right there in his time, but at the timeless flow of stump and Bear and Blueberry. For Greg, the track has just come alive. I watch his eyes lose their focus on the now as he sees the continuum of paw affecting stump affecting Blueberry. The envelope has opened and Greg has entered the borderless realm of past, present, and future.

Crossing an old logging road, we enter a deep, dark Hemlock grove overshadowed by ancient White Pines half again taller than the Hemlocks. The carpet we walk on is soft with the accumulated needles of many winters.

The soil is so acidic from the needles that hardly anything green can grow. My college professors called this a sterile forest community. I wonder if they just couldn't hear the munching of needle-eating Caterpillars and cone-eating Squirrels and Fungi-eating Deer and wood-eating Grubs and Caterpillar-and-Grub-and-Squirrel-eating Birds. Maybe they didn't notice that Deer seek out sheltered groves like this for a quiet place to die. Surely they must have seen Raven, Fox, Fly, Carrion Beetle, Wolf, and Bear coming to feast on the remains. Rodents chew on the bones and antlers for the minerals and to sharpen their teeth. And then there are the Birds who gather the hair to line their nests.

If not the animals themselves, there is the bountiful sign, such as sweet Aphid urine raining down from the treetops to gloss Fern leaves, scat littered along Deer trails, and Owl pellets. Of course, this is just my idea of richness.

"Look at this," says Greg. "I know Deer scat—it is all over my woods at home—and this is on a Deer trail, but this stuff is odd: it is in a little heap. Is it Bear?"

He knows I won't give him an answer—he is just thinking out loud. It's all the more reason to help him think deeper, so I motion for him to remain silent. The more we talk, the more we function on the surface and the less we are connected with our inner selves. This is an ideal setup for Greg to make a step toward that connection, so I want this to be as much a deep sensory experience as possible. If he can get a real feel for the texture and shape and smell of this scat, it could help awaken his inner predator. Without the deep connection, the experience would likely be rationally processed and retained as a mere memory.

Even though jokes are made about "scatology," the study of scat is a legitimate and ancient science. Traditional Chinese doctors will study the scat of their patients as a diagnostic aid. It can show what foods were eaten and how much, how well the food was chewed, digestive speed and efficiency, the composition of intestinal flora, the quality of stomach acids and bile, and signs of stresses, diseases, and parasitic infestations. All of that in a pile of dung! And why not? If we are what we eat, we are what we excrete.

So if you want to find out about somebody, I'd suggest a scatologist over a fortune teller. And for the tracker, the story contained within scat is a good example of an envelope: a track within a track.

I point to a stick and motion for Greg to dig into the scat and see what he finds. He takes to it right away and I'm glad he does because I want him to know something about this animal before we go any further.

If I were tracking this animal, the smell of her scat alone might tell me her story—that she is barely hanging on to life—and for that I would hardly have to break my stride. I would probably come back after dusk to help her release her life-spirit, and to accept her ultimate gift—her body—if someone needed the hide or anything else. Today I am with an initiate who thirsts to also read her story without having to break his stride, so Greg and I will stop and study this scat. Maybe next time he'll be able to keep moving.

Greg shakes his head as though there is something wrong with the animal. Pinching his nose to silently convey to me that the scat smells sharp, he slides his stick over it to show me how slimy it is. He points to a wad of plant fiber he isolated from the goo, and then he uses his hands to suggest a set of antlers.

I nod in agreement.

This Deer is very old and loaded with parasites. The stringy fiber in her scat shows that her teeth are worn down, so she's not able to chew her food enough to digest it well. Her scat is wet and pudding-like, when at this time of year it should be firming up and forming pellets. The scat should be lubricated with clear, odorless mucous rather than foul-smelling ferment.

When food is poorly digested, it enters the colon rich in nutrients. The intestinal flora adjusts in an effort to extract what nourishment it can; however, it usually ends up fermenting and becoming food for worms. Anaerobic decomposition produces gas, which can be told when scat lays in a ragged line rather than a neat pile. Sometimes tell-tale bubbles can be found in the scat, or its surface shows the craters of escaping bubbles. Having a high mucous content from an irritated digestive system, this type of scat usually forms a skin that discolors on contact with air. The thickness of the skin and degree of discoloration, along with changes in smell, are helpful aids in aging the scat, and in the right conditions they can be picked up in passing.

If Greg continues as he is, he'll learn this in time. Right now his thirst drives him to grow in awareness, and rightfully so. One needs a basket before going out to gather.

I gesture for Greg to stand up, and with my extended arm I encourage him to gain perspective. To his obvious surprise, he sees five or six more scat piles within as many paces. I encouraged him to become acquainted with this animal through the first scat so he'd be able to make some sense of her elimination pattern. He'll now leave this experience with the awareness that some animals have elimination patterns and reasons for them.

For this elder Deer, the primary reason for her pattern appears to be security. She is weak from inadequate nourishment and in pain from being bloated, which makes her feel skittish and vulnerable. In the Hemlock grove she feels less exposed and more secure, so she pauses for a moment and relaxes, then allows herself an often-noisy elimination.

If the Wolves don't find her, she'll probably survive the rest of the green season. Then the cold will lay her down. As Deer often do, she may come back to this grove, where she feels comfortable, to return her living breath to the Hoop of Life.

When I get too full of myself and feel the need to be ushered back into balance, I'll sometimes come and sit for a while in a place like this where the Deer come to die. It helps me realize that no matter what I think or do, and no matter how others perceive me, we all walk a common journey that leads to the same end in the Hemlock grove.

Silently, Greg and I walk beside the Deer trail so as not to disturb its integrity. We cross a narrow bog into the grove Grandmother Deer passes through every day on the way to her feeding ground. Her trail passes between towering Yellow Birch and Cedar and Pine who know as many centuries as we know decades. Yet they're just a ripple on this pond, for the song of this forest sings of ten millennia. As I behold this grove before me, ballads of Trees long gone bring grove upon grove to my envisionment. The ages roll before me like a high-speed film, with Ancients rising from saplings and then falling, sometimes alone, sometimes together with branches entwined. Once by wind, another time by fire, and then by disease. Yet every time a new grove rises from the rotting bones, they only fall again, as though they're partaking in an endless dance, bowing and curtsying to each other as one group leaves the dance floor and the other comes on.

Greg drifts out among the giants, his consciousness let loose in the eternity of this hallowed space beneath the lofty canopy. He wanders amongst them for a good share of the afternoon, spending time with this one and that as though he were a child calling upon his Elders.

It is not my place to tell him that he is also an Elder meandering amongst children. He'll come to know this one day when the song that tracks through him resonates harmoniously with the song that tracks through this grove. His ears are only opening now; the time will come for them to hear so much more.

We sit upon a mound between two Pines; one is dead, the other alive. Each of them is big enough to supply enough wood to build a house. I motion for Greg to notice the contour of the mound, which is similar to the Spruce mound from earlier in the day but much larger. His frown tells me he doesn't make the connection.

Gesturing for him to follow me, I take him to a mound on the other side of the dead Tree. Like the stem of a capital T, it has a long, low mossy hump extending from it. Atop the hump grows a line of young Hemlocks. Except for them, the nearby forest floor is bare. I motion for Greg to dig into the hump for a handful of soil, and then into the forest floor for the same. In one hand he holds dark, moist humus from the hump, and in the other is dry, gravelly sand.

Continuing to honor our silence, he raises his arm and brings it down slowly to show a great Tree toppling to its final rest. With the upsweep of his hands, he shows the young Hemlocks growing up from the rotting hulk. Greg looks out over the sturdy saplings, and then up to the spires of their grandparents, and he drifts away to a place that only he can know.

When he returns, I direct his gaze with the sweep of my hand from the mounds before us on to the panorama of mounds that extend in every direction as far as we can see. No longer do they appear random and scattered, but we see their parallel formation across the forest floor, like the crests of wind-whipped waves. They come alive, slowly rising and falling over the centuries like swells in a sea of Trees.

Greg holds up his arms and rocks along with Trees being torn at by a vicious wind. A tremendous gust screams through the canopy and in one terrible moment the roots of a thousand forest monarchs are ripped to shreds. A majestic stand that toiled for an eon to reach up into the realm of the Sun Father comes crashing back to the bosom from which it was born. An era is ended.

And our day is done.

4
Dogs Will Be Dogs

Learning through Conflict in a
Wayward Track across the Snow

Every type of roof has its own way of shedding water, and by listening one can tell the kind of roof, the intensity of the rain or snow melt, and sometimes even the temperature and time of day. In a similar way, Pine, Oak, Birch, and Maple each sing differently in the wind and tell their stories about the weather and the season.

The distinctive sound of water dripping off our cabin's cedar-shaked roof woke me up this morning and reminded me about Birds and Insects teaching me how they speak with their talking wings. Each species makes a unique sound when flying, and it is even possible to identify individuals. Not only are they going from point A to point B, but the melody of their wings tells anyone who cares to listen to their age and gender, how they're feeling, and what they're up to.

I remember in particular the time when I'm out in the woods and I hear a ruckus from a family of Sapsuckers. It grows more intense the closer I get to them. I want to listen with my full attention so I can better "see" what is out of sight, so I close my eyes and stalk in closer. Right away sounds grow sharper, and I can more easily tell what direction they are coming from and from how far away.

I've come to trust that when one of my senses is compromised, the others compensate. Even with one sense so radically different from another, such as sight and hearing, either one can take up some of the slack when the other is down. I have a friend who is hard of hearing; however, his sight is so sharp that he often catches things I miss. The same thing happens to my hearing as soon as I shut my eyes while approaching the Sapsuckers. Sounds seem to come out of nowhere that weren't there just a bit ago.

I don't want to create any more disturbance than I already have, so I approach them off to one side to let them think I am just passing through. Their calls tell me they have relaxed a bit as I go by them, and I take advantage of their attention lapse by transforming into a stump. As I envision my flared roots, peeling bark, and jagged top, my breathing slows and my sense of self fades. I let my ears tell me the story of what is going on around me.

On my exposed and vulnerable bark, two Mosquitoes and a Deerfly let their culinary intentions be known, and a couple of body lengths ahead a Woodcock grumbles. I guess they haven't entirely bought into my stumpness. A mother Robin, about thirty paces ahead and off to my right, lets everyone know how agitated she is by the carrying-on of the Sapsuckers. She is likely concerned that drawing attention to anywhere near her nest will endanger her young.

Trees drip from the shower that just came through, and I listen intently to the droplets hitting leaves, needles, branches, and ground. *Pftt* says "Maple leaf," and the *pftts* draw me a picture of a dense cluster of Maple saplings immediately behind me. *Tih, tih, tih,* quick and light, trace a clump of Hazel brush less than a body length off of my left shoulder and slightly back. The wall of barely-audible splats just to the front and left, along with the light mist showering me, tells me I am next to a conifer. Considering the habitat, she is probably a Balsam Fir, and the drips hitting her are likely coming from a large deciduous Tree canopied over her. The steady rain of drips in front of me indicates a grove of Aspen or Birch (I might be able to tell which if there was a breeze, as Aspen leaves flutter, which dries them faster than Birch's). The groundcover sounds to be a patchwork of Sphagnum Moss, leaf litter, and Bunchberry. A visual diorama painted by sound.

Now that I feel comfortable knowing the surroundings, I pick up again on the Sapsuckers. The male's solid, steady wingbeats speak his seriousness and determination as he takes a sentinel position overlooking the area. He is deliberately quiet when switching positions to gain more perspective. I suspect it is so he won't detract from his mate's attention-getting antics. Her flight is frantic, nervous. She stays low, hovering around her young to draw attention away from them and toward her. The little ones' rapid wingbeats and short flights often end in crash landings, telling me they are just out of the nest and can't get up in a Tree where they'd be safer—and where their parents could be more relaxed.

The memory fades as the dripping roof gets louder. The previous two days gave us eighteen inches of snow, and if it were late autumn we'd have a good start on a fluffy insulating blanket for the white season. However, it is the first of April, which here in the Northcountry means that it could either be an April Fools' joke and disappear in a day, or it could stay with us for a few weeks. Judging by the brightness of the dawn sun and the intensity of those drips in what is usually the coldest part of the day, I place my money on April Fools.

Shortly after Lety and I arise, Justin, a student with a passion for tracking, appears at the door. He is a mostly serious, philosophical type who comes in from the students' wilderness camp only when he has to.

"I . . . I hate to disturb you two," he says in his usual self-effacing way. "I don't know if you're busy . . . I apologize if I'm disrupting something." And then with bright, dancing eyes, he blurts out, "There are some neat tracks just outside in the yard—they're all over the place! They crossed over the pond, went around the log cabin, and then over to the outhouse."

"Sounds like a tracking adventure," I reply with mock overexcitement. It's not that I'm not curious about who was wandering around in the night; I just like to ham it up with Justin. I do it partly because he tolerates my humor, and partly in payback for me tolerating his. Lety and I are immediately out the door and following Justin's lead without even thinking about grabbing coats.

As we approach the area of activity, I start to hear the song of the animal's track. Envisioning myself coming into the yard just before first light, I put my nose to the ground and sample the scents that hang heavy on the damp, warm air that is pushing up ever-so-gently from the south. Inhaling deeply, I luxuriate in the intoxicating overload—such a treat after a season of cold, thin air that doesn't carry scent very well.

An envisionment for me is like watching a video, while at the same time I'm in it, living it. This envisionment gives me my tracking experience, and it is all I need in order to know the animal's identity. I am pretty sure of his sex, what attracted him to the yard, the direction from which he came, and the direction he was heading; and I have a hunch as to what he checked out first in the yard and what he headed for after that.

I am ready to move on, or I should say "go backward," as I like backtracking. I learn more about animals and people—about who they are and what they're up to—and I learn it faster, by finding out where they've come from rather than where they're going.

However, my tracking partners are still busy learning from what is right in front of them, so I have time to kill. I decide to listen more to the song of the track, to see what else it might tell me. I'll study the actual sign, such as the tracks, scent markings, and whatever else the animal left behind, later along with Lety and Justin. It would augment what the song of the track gives me.

One voice of the song that comes through strongly is from the weather. Some of it drifts in from past experience, some from yesterday and last night, some from what right now touches my face and tickles my nose, and some of it just wells up from a deep place of knowing. I don't have to think about the direction of air flow, how the snow was melting in the night, or the lay of the land and how the essences that attracted this animal were flowing over it. These things just subconsciously contribute to my knowing the animal's motivations and movements.

I don't want to steal Justin's fire, and I know it could be a more complete experience for him if I remain silent about what I am picking up. In order for him to know the song of the track, he needs to learn how to listen. If I tell him details about the show before he sees it, what would it do to his sense of adventure? How could he experience the joy of discovery? His mind would wander off and he'd miss so much of what the experience has to offer.

On the other hand, when there is a mystery to be solved, a state of inquisitiveness infuses the entire body. A moderate flow of adrenaline sharpens reflexes, senses become super keen, and the mind buzzes with activity. Memory banks are scoured for any bits of information that might help make sense of clues, and free associations are made to try and figure out the plot by piecing various scenes together. It is just as if we dropped in late on a movie (and with tracking, one is always showing up in the middle of something). We'd try to figure out what we missed, and then we'd start working on what was going to happen.

As helpful as the rational mind can be, it can also limit one's tracking ability. It specializes in A + B = C problems, and the track of an animal—not to mention a human's—has so much more to tell than fits into a simple formula. Imagine someone went through the kitchen and left footprints. It is obvious he passed through, but is that the real story? Where was he coming from and what was he after? Was he empty handed or carrying something? Did he drop anything off or grab anything? Was he in a hurry or preoccupied? What was his mood? What did he notice and how did it affect him? Where did he go after the kitchen, and why? This is the story that is often missed in tracking. It's not just the fact that he went from one end of the kitchen to the other.

When we have a way of listening to the story—and we do with intuition—we have a way of answering the questions I just posed, and we don't always have to see the footprints. Besides, relying on them is like reading the sheet music rather than going to the concert and catching the energy of the live performance.

Of course, it takes some tracking ability to hear the echo of the band's performance, as opposed to going straight for the sheet music. It's like a paw print in that it is a known entity that can be sketched, photographed, or plaster casted. The song of that animal's track, on the other hand, is ethereal. Hearing it involves your whole being. It takes becoming the animal and experiencing his reality.

I reconnect with Justin and Lety. Justin, who approaches tracking like a religion, had already shadowed the animal's movements by following his trail all around the yard. He kept a half arm's length beside the tracks so as not to disturb them. From where we stand, he points out where the trail goes, which is close to what I envisioned. The animal came up the driveway from the road. There is dense forest on the right, which opens into a small clearing about fifty paces up. Lety's and my house is on one side of the clearing, and on the other side is a small pond that is still frozen over. The animal turned toward the pond as soon as he saw it. To the left of the pond is a log cabin, and he followed the shoreline toward it.

When he reached the cabin, he turned left to head back to the driveway. He took only a few steps, then turned around and went in the opposite direction, following the path that runs between the cabin and the pond to the outhouse. Walking up to the outhouse door, he took a whiff and did a quick about-face. I'd have done the same. Give me the fresh outdoors any day to sitting in a smelly little box over a cesspit. He backtracked to the pond, crossed it, and cut through the brush to a two-rut lane that took him back down to the road he came from.

"What do you think?" I ask Justin.

"Well . . . I think Coyote. Maybe he has moved into a new territory and is exploring around, or maybe it has something to do with the time of year. . . ."

"What did the song of the track tell you?"

"Well, I don't know. But I see that he is a pretty curious puppy. Either that or he has a short attention span. He went to check out the pond right away, then it looked as though he was gonna go around to the front side of the log cabin, but instead he walked over to the outhouse. Then he came back over to the cabin and changed his mind again and trotted across the pond and scooted down the sand road instead."

"How does that compare with other Coyote tracks you've seen?"

"Good point. They're usually more direct—they seem to know where they're going and what they're up to."

"How about the size of the pawprints?"

"It looks too big for a Red Fox, and maybe for a Coyote too. I've seen Wolf prints and they're as big as the palm of my hand, so I know it can't be a Wolf, unless it's a small one."

Fox's footprints can often be told from Coyote's at a glance, regardless of size. Coyote's is rounder than Fox's, and Fox's usually shows short outer toes that reach just up to the middle toes. This is true for all of our North American Foxes, whether they're Red, Gray, Arctic, or Kit. Coyote's outer toes, on the other hand, are typically longer—they wrap up around the middle two.

It gets a little more confusing with Dogs and Wolves, as most Dogs have a toe-length configuration similar to Coyote's. Notice that I said *most* Dogs. They come in so many shapes and sizes that generalizing anything about them would be like claiming that all Birds have short tails. And then there is Wolf, whose toe length you'd figure would be like Coyote's because they're closely related. But oddly enough, Wolf has toes like Fox's.*

Jason doesn't know this, and too much technical detail right now might just confuse him, so I encourage him to stay with the track he is on by asking, "Are you sure this is a canine track?"

He puts on his signature serious look, takes a deep breath, and squats next to the track.

"Pretty sure," he responds after some studying. "Look here at this clear footprint. It's the back foot—it's smaller—and there's the canine X formed by the space between the pad and the toes. And see the little pyramid of snow right where the two lines of the X cross? And look how the side toes are tucked in neatly behind the two middle

---

*For more on identifying canine tracks, see appendix 2.

toes. A cat's side toes would be more dominant and spread out."

"Well then, if he's a canine for sure, put all you now know together. What brand of canine do you see?"

"Oh, I get it!" He exclaims after a short pause. "A Dog! No, it couldn't be. A Dog has a wider straddle than that, and the toes are more spread out, more puppy-like. If this is a Dog he has a long stride for his size like a Coyote or Wolf does, and the toes are pulled in together, especially those side toes, just like I'd expect to see with a Coyote track or a Wolf track. But now that you mention it, this critter's acting like a Dog, like an animal just let out of a cage. He's all over the place, full of energy and curiosity."

"There are Dogs, and then there are Dogs," I say. "Envision the chest of a Lab, and then a German Shepherd."

"Right!" Justin replies immediately, as he knows Dogs well. "The Lab's is broad, more like a puppy. She'd have a wider straddle and a sloppier-looking walk. I bet her footprint would look more puppy-like, too. The paw would be spread out, and without those two center toes being so long and dominant. The Shepherd is lankier, more wolf-like actually, so I'd guess his track would look more wolf-like."

Shepherds are more recently evolved from Wolves than are most other Dogs, so even though a Shepherd's temperament and mannerisms are very Dog-like, they still show much of their Wolf ancestry in their physical structure. Hunting Dogs, on the other hand, have been domesticated probably the longest of all Dogs, so they least resemble Wolves. They're the most neotenous of all Dogs, which means they retain puppy-like characteristics as adults. This includes a yearning to play and socialize, the ability to stay focused, blunt, rounded features, droopy ears, and big paws. Even many toy Dogs, who seem so far removed from Wolves, have a more audacious, independent character than hunting Dogs.

"No wonder I was confused," says Justin. "The footprints themselves looked Coyote-like, and I was focusing on that, rather than stepping back and gaining perspective. Now I see it. Look how this critter was running around here, this way and that and this way and that again. That's the last thing a Coyote would do."

"Let me add to your confusion," I kindly offer. "You're talking about erratic behavior; do you know much about Wolf-Dog crosses? I'll tell you, those poor things can be schizoid! Our neighbor up the road has one, and if these are his tracks, it might help explain why they look Wolf or Coyote-like to you. It's hard enough to distinguish Dogs from Wolves or Coyotes sometimes—especially for beginning trackers—and hybrids can easily confuse even experienced trackers. Here's a time when backing off and listening to the song of the track can be very helpful."

Crossing a Wolf with a Dog is like trying to mix oil and water. Rather than an animal with blended traits, what you end up with is a canine Frankenstein—a beast with a bright, wild spirit struggling to be free of the placid domesticity clouding his mind. The result is a psychotic creature so tormented and unpredictable that out of the blue he could lash out at anything or anyone around him. Sometimes it is furniture or other pets, and sometimes it is children. He might be the most shy and gentle pet for years, and then one day he just snaps.

These unfortunate misfits are like some people I know, whose primal selves have come alive. They feel either trapped by responsibility or crippled by lack of courage. Some of them snap, just like a Wolf-Dog. And some of them own Wolf-Dogs. I wonder if it is because they feel an innate sense of kinship.

"And hey," I add, "if you think you have a challenge here, imagine if you were in Australia. Sometimes stepping outside of our world and looking at a parallel can help bring us clarity. Australia has introduced domestic Dogs and Red Foxes, just like here. They have imported American Wolf-Dogs also, believe it or not. And on top of that, they have the Dingo, who seems to be neither Wolf nor Dog, but something in between—and for quite a different reason than our Wolf-Dog crosses."

I go on to explain that the only canine in Australia when the

Europeans "discovered" it was the Dingo. Even though she lives in a wild state, she is not a true wild animal, nor is she native to Australia. I believe all the Dingo traces archaeologists have found are less than five thousand years old. She is related to Southeast Asian Dogs, so she was probably brought over by seafarers.

With her broad chest, rust-colored coat, and upturned tail, you'd think she was a Dog. But then she has the pack social structure of a Wolf, and like a Wolf she goes into heat only once a year. Even her skull shape falls somewhere between Dog and Wolf. And the oddest thing is that she doesn't bark.

I don't think the Dingo could have survived if there were already wild canines in Australia. With no competition, she had easy living. Wolves have an easy solution for canine competition: make a meal of them. In fact, the only way Dingo-like Dogs are able to survive in Southeast Asia, where there are also Wolves, is to hang around human settlements and scavenge garbage.

"The Dingo," I continue, "along with Australia's hodgepodge canine situation, is similar to what we have right here. If you can draw the parallel, it could help you with this last variable we have to add to the jumble. And if you don't have a headache yet, this should do it—we have another type of Dog with an odd gait. . . ."

Justin, who has studied Dog gaits, is on to this one right away. "You mean like wiener Dogs, those short-legged things with the long bodies? Yeah, I've watched them run—their rear end swings way off to one side, so their back feet aren't in line with their front feet. The track looks so weird, like two Dogs running close together."

I tell Justin and Lety about the time I spotted these tracks up ahead on a Lake Michigan beach. It looked like two animals running tight beside each other in perfect step. They looked so uniquely beautiful in the distance—and did I have a good laugh when I got closer and realized it was just a Dog's back end off center from her front. Sometimes when I see long, low-slung dogs try to run, I picture what a Weasel, with his typical Slinky-toy bound, would look like if he tried to side lope like a Dog. A side lope, by the way, is the track pattern Jason is referring to. When you see a Dog cruising in an easy rocking-horse rhythm, with the front feet coming down one well before the other, and then the two back feet following suit, she is probably loping.

Lety, Justin, and I go on to talk about how tracking Dogs can be just as challenging, if not more so, than tracking wild animals. They're just the ticket for tracking exercises. They're around nearly anywhere and anytime, their sign is easy to find, and they're obliging. If you want to see how a track looks in wet sand or what a trail through tall Grass looks like, a Dog will be glad to help.

Tracking Wolf or Cougar might have more romantic appeal than following the family pet around, yet that mutt's tutelage could make it possible for you to someday track those elusive shadows of the wilderness. It has worked well for me, and I got an added bonus: it has helped me find and read human sign, along with understanding the habits behind the sign. As Jason is discovering, Dogs, even though descended directly from Wolves, exhibit un-wolflike behaviors. The same is true of us when compared with our hunter-gatherer Ancestors. In fact, modern Dogs and modern humans hold some behavioral characteristics in common. How this came to be is a mystery that has fascinated me to no end. My theory is that we wanted to more easily live with Dogs while they were doing our bidding as hunting Dogs, racing Dogs, lap Dogs, show Dogs, and so on, so we selectively bred them to have humanlike qualities.

In the process, we've created an animal who can be bought and sold, and who will serve her master, often whether he is right or wrong, kind or cruel. I've seen too many Dogs given substandard food and shelter, along with being chained and locked up like captives or slaves.

When I apply human relationship standards to some of our relationships with Dogs, I question their healthiness. When a human keeps coming back to a controlling or abusive relationship with a smile, most of us will recognize it as dysfunctional. However, when we witness the same behavior in a Dog, we often call it loyalty, or even love. I've seen it over and over in Dogs who are chained, ignored, or mistreated, and they just keep coming back for more—often with tail wagging.

In humans, this behavior would be called *codependency*, which is the neglect or denial of self to please others, in order to maintain the relationship. In Dogs, the capacity for codependent behavior is likely a selectively bred trait, as I haven't seen it in Wolves, their forebears. Wolves will either shut down or attack, but they will seldom kowtow—at least not for long.

I know people, myself included, who struggle or have struggled to have fulfilling human relationships, or who are loners for various reasons, and turn to Dogs to meet their needs. Many Dogs will provide touch and companionship without asking any questions.

Perhaps the better we come to know Dogs, the better we can come to know ourselves. I've found that the life of some Dogs can be a fitting metaphor for those of us who feel controlled and unfulfilled. I remember times when Zhavago, my sled Dog buddy, would run off on wild adventures across the countryside, and I'd have to track him down and drag him back home. My truth was that I wanted to take off with him, but I'd stuff my frustration and do what was expected of me. He wasn't being disobedient; he was just more emotionally honest than me, which shamed me when I hid my feelings and tried to dance around my pain.

Here is where listening to the song of the track can give us so much. When we approach tracking as mainly the study of the pawprint, we tend to remain detached from the animal. On the other hand, when we listen to the song of his track—the feelings and motivations that moved him to act and react the way he did—we're hearing some of the song of our own track. We're given a window into the yearnings, fears, and neuroses of the controlled life we call domestication, whether it be canine or human.

For a time back in my early adulthood, I was hung up on one

characteristic of domestication: lack of awareness. When out in the woods, I had things on my mind, so I couldn't be fully present. I was clumsy, I made noise, and I missed things I should have picked up on. The odd thing is that I'd get irritated by my Dogs doing the same thing, but I couldn't see it in myself.

One thing that bugged me was where Dogs pee, which is wherever they happen to be at the moment. This is truer of confined Dogs than those given some freedom. I expected them to act more like their wild cousins, who chose their pee spots to mark territory, signal sexual receptivity, or challenge rivals. Wolf scent posts could be boulders, Tree trunks, tufts of Grass, or scraped-up mounds of dirt, and they are periodically remarked. Oftentimes somebody else's scent would be marked over, to let the locals know who passed by. And it wasn't just males—some females marked as well.

One day when I was out with my Dogs and had to pee, it dawned on me that I was doing the same thing as some of them. I'd just take a leak unconsciously, in the most convenient spot, which could be alongside the trail, a step outside of camp, or right where I was standing. I felt shame for judging them while being so blind to myself, and at the same time I was grateful to them for mirroring me so that I *could* see myself.

Like wild canines, there are aboriginal humans who deliberately choose where they pee. Some maintain scent posts to keep animals from coming close to camp. I've also learned to pee consciously. I now avoid reusing the same spot, to reduce odors and keep from killing plants. Regularly used sites can get torn up by animals attracted to the concentrated salts. And I relieve myself away from water, to keep from contaminating it.

How we pee is a good example of how I see degree of awareness as the main difference between our Ancestors and us, as well as between Wolf and Dog. As Justin concluded from this tracking experience, Dogs will be Dogs, and that is all they can be. As the Dingo seems to show,

Dogs' Wolf origins are largely bred out of them and they can only partially return to their ancestral form.

Fortunately, such does not appear to be the case for us. As much as we might be like Dogs in some ways, our wild side is alive and well. Genetically, we've changed insignificantly since our hunter-gatherer days. We're perfectly capable of living that way again, which includes being superb intuitive trackers.

As for Dogs, I'd suggest that rather than hanging on to our romantic notion of the noble Dog as man's best friend, we see him for who he is: a creature of our creation, an incomplete Wolf who finds wholeness in his association with humans. Why change our view? Because a tracker needs to cut through illusion and belief in order to see the track for what it is. And because a Dog has the most to offer by just being who he is. If we can see this, perhaps his example can help us just be who we are.

5
Stalking Turtle

Now Step Back and
See Life and Death in a Track

The track of the Drought lays heavy upon the land. Its song wails like a funeral dirge. Lakes and springs dry up, moisture-loving Trees like Birch succumb, and wildfires rip through forests and meadows that haven't been burned for centuries.

Paradoxically, the air hangs heavy with moisture. Yet it doesn't get sucked up in the thermals of sweltering air that rise and irritate the Thunder Beings, which usually causes them to retaliate with drenching downpours. At this time of year, the Thunder Beings normally replenish our waters with an inch of rain a week, and we're barely getting an inch a month.

Two days ago, I stood on a Beaver dam and let my eyes travel down over the dry stream bed. It wasn't enough—I needed to feel the crusted muck with my feet and hear the crunch of Snail shells and baked pond weeds. I walked down the stream bed and came upon two streamside Tamarack Trees who had been flooded out a couple of years ago by the backed-up waters of another Beaver dam further downstream. What a contrast! Death from too much water right beside death from too little.

Tucked behind one of the trunks was a tunnel into the bank. It was the entrance to a Beaver den. At one time safely underwater, the entrance was now exposed, which forced the family to move. I found them upstream on a large headwaters pond that still had plenty of water.

The deep holes in the stream where the last of the water stood were lined with dead Fish and Tadpoles dehydrating in the hot sun. From a distance, it looked like a scene from a courtly legend. The holes, glazed with baked scales, glistened like large silver-lined bowls waiting to be filled with fresh water to quench the thirst of the royal steeds.

Discovering the entire stream bed to be dry, I returned upstream to the base of the dam, where the last remaining pool of water was slowly shrinking away. To call it a pool of water was generous, as it was more of a muddy gruel teeming with Minnows and Tadpoles gasping for air. In answer to some primal voice, some of them burrowed into the cool muck at the edge of the pool, hoping it would keep them alive until the rains returned. Their forebears who did so lived to pass the memory on, and the ones who didn't suffocated and baked like those who made up the shiny lining of the death pools downstream.

The song of the Drought's track around this little mud hole was so blaring that I wanted to shut it out. Instead I closed my eyes and let the song carry me back to the previous evening. A mother Raccoon and her two kits came down early to the pool for an easy feast. The family, the only one I know of in the nearby area, came up the creek bed looking for stranded Crayfish, Leeches, or anything else to please their omnivorous palates.

Usually the more shoreline, the more Raccoons. We have miles and miles of stream, pond, and bog shoreline, yet only a few Raccoons. Even though their range extends right up through our area into southern Canada, our acidic bogs and piney woods aren't their preferred habitat. What they really like is deciduous woods with sweet wetlands.

I watched a giant Water Bug suck a Tadpole dry. Out of the grassy bank at my right, the head and neck of a Mink emerged. She sniffed the air and ground to the right and left so conspicuously that it looked as though she was nearsighted and straining to see. In her closed-in world where she often doesn't have the clearance to see any distance, keen senses of smell and hearing are vitally important. Slithering straight down to the pool, she knew exactly what was there awaiting her. She was quick as lightning, even though her quarry wasn't going anywhere. Mink have only one speed; it's just the way they're built.

Opening my eyes, so to speak, I returned to the present. My eyes weren't literally closed; I just tuned out the day in order to tune into the scene around the pool before dawn. It is a necessary step in preparing to envision.

Contrary to the beliefs of managers who want to increase productivity and those who think they can watch a video, text message, and carry on a conversation at the same time, we are not designed to multi-task. Efficiency, accuracy, and recall decrease dramatically when we try to do more than one thing at a time.

When the consummate tracker is tracking, she enters another world.

Everything else ceases to exist, including her own consciousness. It is as though she is a separate animal from who she is when she is not tracking. She is like a hunting Mink, designed to operate at one speed, for only one purpose.

Coming back to the reality before me, I was immediately drawn to the white blotches freckling the drying mud around the edge of the puddle. It looked like the scat of the Plover I spooked when I first came over the dam. From the number of splats, I figured he had been hanging out here and feasting on the easy pickings for several days.

Hmm . . . I was finally getting it—Raccoon, Mink, Water Bug, and now Plover, were clearly telling me there was more to this Drought than the grim picture I first saw. It was a good lesson for me in stepping out of disaster mentality and into the greater awareness of the Hoop of Life.

Even so, I felt for the trapped victims. I dug a little trench through the dam to allow enough water through to fill the deathtrap immediately below. I had to be careful with how much water I took lest I jeopardize the pond above the dam.

The Fish responded instantly upon feeling the flow of fresh, cool water by wriggling out of the mud and swimming en masse up my little trench to the water's source. What a sight to see! They piled on top of one another in their effort to escape their mucky mass grave. I sat there on the dam for a good share of the morning, watching the pool ever-so-slowly fill. I felt such joy as the compressed life of the stream spread out to recolonize their newly expanded world. They acted as though this was planned all along, they were well rehearsed, and they were just waiting for the opportunity. And of course they were, because they've been practicing at this ever since droughts have come and gone.

A Frog hopped down the bank and into the water, and the Plover kept flying over. Each time he tried to land, he was startled off. He got used to me being there, yet he couldn't seem to accept that his favorite mud flat was fast disappearing.

I imagined how fast his dried scat, which I now saw through a lens of water, would dissolve and become food for Plankton, who in turn would feed Tadpoles and Fish, who in their turn would nourish the Plover. In the grand scheme of things, drought seems to be only a momentary disturbance, hardly disrupting the Hoop of Life. In fact, as I reflected, I observed right before me this example of how the Drought actually nourished the Hoop.

Yet I chose to interfere. Someone asked me why, and I didn't have a ready answer. My usual approach is to let things unfold as they will. Greater purpose, minimal impact, nature's way, non-intervention. Still, I've rescued drowning Birds and Dragonflies, and I've killed a number of injured and suffering animals. I may have done it out of empathy, which is a trait common to many social animals. Wolves have adopted abandoned human babies, Dolphins have saved drowning sailors, and Leopards— even though not social animals—have protected orphaned Baboons. Another factor was the sense of responsibility I feel for human-caused environmental destruction. Parts of the stream and surrounding area were altered by human activity, which magnified the Drought's effect.

It is now a little past sunrise and I'm coming down to check out the pool. It has been a couple of days since I breached the dam and I want to adjust the flow rate. Just as I step onto the dam, I hear a disturbance at the far end of the pool. Right away I think, "Plover," but no Bird flies up. Instead something splashes down into the water—something big— but I can't make it out. Ripples on the surface say it is heading my way.

A blocky head emerges, followed by the top of a dark, plated shell that looks to be the size of a serving platter. The Fish panic—they shoot off in every direction from the creature, skimming along the surface like skipping pebbles. But they can't go far, with the pond being only about six paces long and two wide. They run into the bank, then tear off in another direction. They must have an instinctive fear—no, make that terror—of Snapping Turtles, as their reaction to his presence in the pool is more extreme—far more extreme—than their reaction to mine.

The Turtle plods steadily up the length of the pool, stopping every few feet to raise her head and attune to the surroundings. As nearsighted as Turtles are, I doubt that she sees me. If I were moving, she'd know my presence by vibration, which Turtles are very sensitive to.

I know not to freeze in place, as that would also cause a vibration. To be rigid in a world of movement can cause as much disturbance as moving faster than the surroundings. I let the wind rock me slightly, just as it would the skeleton of a Tree drowned by the dam. So she won't feel me focused on her, I let my senses drift into overall awareness mode. My mind drifts and I pick up on anything and lock on nothing.

By the look of the dried Algae on the Turtle's shell, I figure she has just made a long trek up the dry stream bed and is anxious to soak herself in a big bathtub. For sure, this little puddle won't do—it hardly covers her back. Like the Fish, she keys into the fresh flow of water coming into the pool. Unlike the Fish, she is much too big to squirm into the little trench I had dug, so with powerful forefeet and gripping claws, she pulls her hulk up the face of the dam. Efficiently, but not very gracefully, she maneuvers around, over, and under protruding sticks.

When she reaches the top of the dam, it looks as though she is going to crawl right over my feet. About a forearm's length away from me, she senses that the nearest water is over to my left, so she veers that way, never paying me any heed. She plops unceremoniously into the water and submerges herself with all but the top of her shell covered. Remaining still for a long moment, perhaps to adjust to the cold water, she then disappears under the floating vegetation that rings the shoreline and swims out into the pond.

As I watch for her head to break the surface, an adventure with one of her relatives comes to mind. It was early in the afternoon on a cool, sunny day late in the Strawberry Moon, which falls around mid-July. I was canoeing, my favorite mode of transportation, and I was out on what the Ojibwe call *Mashko-Sipi:* the Grassy River. Its headwaters are a great, open bog, through which the fledgling stream lazily meanders—perfect for my lazy, meandering paddling style.

The bog that births the Grassy River is nestled in the middle of an enchanted wilderness that sits right atop the eastern continental divide. Its waters drain both westward into the Mississippi River and eastward into the Atlantic Ocean. A unique feature of this wilderness is that even though it sits high on the divide, it is a rather flat plateau—very unlike the jagged mountains along the western continental divide.

The levelness of the plateau makes it a water wonderland. It is a forty-by-sixty-mile area with over three thousand lakes and countless streams, rivers, ponds, and bogs. Someone with a yen for algebra figured that 40 percent of the plateau's surface area is covered by water.

I cruised along with little effort, as most of this water is tranquil enough to be easily paddled, and for me that spells paradise. From there I could have gone virtually anywhere I needed to by water, including Wild Rice beds, prime trapping and fishing areas, berry patches, Maple groves for sugaring, Birch, Cedar, and Spruce stands for craft materials, Sedge meadows for thatch, stands of primeval forest for reflective time, and the list goes on.

Leaving the bog, the little stream gradually became a river as springs and small brooks fed into her. I drifted along with the quickening current between drumlins, which are big, oval hills crafted by the glaciers that once scraped their way through this area. The banks now often rose high on one side or the other, and the riverbed was strewn with boulders.

It was now late afternoon and the sun was at my back. Big boulders standing above the water glistened in sun-baked splendor. One particular boulder caught my eye; it had a cap that shone brighter than the others. I drifted closer and saw that it was a Turtle—a big Turtle. I thought he must be a Snapper.

When I was a kid, catching and studying Turtles was one of my favorite activities. They're some of the most amazing sign I've found on the track of evolution. Find a Turtle shell and look inside, and if you don't already know what the shell actually is, I bet you'll be surprised.

A couple hundred million years ago, the reptilian ancestors of Turtles decided to return to the water and live again with their cousins, the Fish. Only these reptiles discovered they had lost some of their ability to function as true cold-water creatures. Even though technically cold-blooded like Fish, they learned a couple of tricks to warm themselves when they lived on land. One of them was lying in the sun. In time, they became dependent on the sun's warmth to help digest their food and give them energy for rapid movement. This is why virtually all reptiles either live in warm climates or spend a lot of time sunning themselves. Sea Turtles, who don't sun themselves, are found only in warm tropical waters. The more northern a fresh-water Turtle, the darker her shell, which makes it a more efficient solar collector than a light-colored shell. Many northern Turtles are small so that they can warm rapidly. The Snapping Turtle is an exception. To compensate for his size, he lives a slower life than smaller Turtles and basks longer in order to warm himself through.

I wondered how long the Turtle up ahead of me had been basking. The song of his track told me he'd been there a long time. His shell was bone-dry and there was no water dripping down the rock. Yet that wasn't reliable sign, as things can dry fast on a sun-baked rock. The fact that it was late in the day wasn't dependable either, even though large Turtles, who take time to warm up, like to start sunning as soon as the rocks are warm. I figured he'd been there awhile because I sensed he had settled into a suspended state I call oneness, in which one can be very relaxed, as though napping, yet maintain sensory alertness. It takes time for a Turtle to relax into this state.

A primal urge welled up from deep within me and took over. There was no shadow to meld into and disguise my movement, so I became the movement around me. My paddle was the current that guided my boat to the rock, my eyes were the breeze that drifted around everything without stalling on anything in particular, my boat was a log drifting down the river.

Whoosh—my hand darted out toward the Turtle like a Snake

striking, and just as quickly my hand turned back to a branch dangling off the log. Inwardly I let out a victory whoop, and the log continued its mindless journey.

A Plains Indian warrior's greatest honor comes from counting coup, which is accomplished by touching the enemy without anyone getting hurt. The touching is the easier part; escaping to tell the story is the real challenge. It is the same challenge a warrior faces when he needs to sneak into a place—once he is there, he has to get back out undetected. The same was true with my counting coup on the Turtle.

I kill only when I need to. Not so with counting coup, and I hardly ever pass up the opportunity. It is excellent training, which helps me stay on top of my game so that I'm ready when the hunt calls. Counting coup is more difficult than hunting because I have to get close enough to touch the animal. So if I can count coup, there is a good chance I can make a kill when necessary. And I can practice counting coup virtually any time or place, as it requires no equipment, it is always open season, and everything is fair game.

Killing is another story when it comes to broad and sweeping events such as droughts. It isn't just life one second and bam—death. And death itself is only a small part of an intricate play that is being acted out. Imagine catching just the death scene of a movie; I bet you'd feel like you were missing a lot.

With a natural "catastrophe" such as a volcano or hurricane, it is quite easy to see how death so quickly transforms into life. Carcasses are scavenged to feed young, and even the uneaten scraps nourish plants and Insects, who in turn become food for other animals. Flattened forests and flood-ravaged bottomlands are quickly blanketed with seedlings racing each other to reach the sky. Forest fires are necessary to help some Tree seeds sprout. Whether it is one death at a time or a million at once, it is all part of the endless hoop that goes from life to death to life again.

6

How to Learn Tracking from One of the Greatest Predators

Apprenticing to the Wolf on the Windowsill

When I start talking about the tracking skills I learned from living with master predators, most people who know me think "Wolf" right away. Without a doubt the family of Wolves who welcomed me into their life when I was a young man gave me intimate knowledge of the hunt. This story, however, is about another type of Wolf, who patiently instructed me in the mystifying ways of the stalk and the kill. And he showed me their terrible beauty.

Unlike Wolf, this is a teacher available to each and every one of us. In fact, he is bound to live somewhere in your part of town. He runs wild in many fields and parks—and possibly even in your backyard. Let me tell you about one day last green season when Stefan, a gung-ho yearlong-program student from Switzerland, met this tracking teacher. I walked into camp and Stefan was wearing a smile that wouldn't quit. "What got into you?" I asked him.

"I'm not sure," he replied, "but I really like it. At dawn, I went up on the ridge by your lodge, and when the sun broke over the horizon it lit up thousands of spider webs sprinkled with dew—they carpeted the bog! I don't think I could have taken a step without destroying one. I can't imagine how many Spiders there must have been to make all those webs."

Few people pay much attention to Spiders, and fewer yet realize just how many of them there are. I know a naturalist who claims that in most environments you're probably not more than an arm's length from a Spider. There are Spiders who travel on the wind like Dandelion seeds, there are Spiders who live underground, and there are Spiders who live underwater. From windblown treetops to the stark faces of skyscrapers, Spiders inhabit virtually every available niche. Some Spiders are active year-round, even here in the North Country, where I see them out and about long after other Bugs have either frozen to death or gone into hibernation. We get temperatures as low as minus thirty degrees Fahrenheit and it doesn't seem to faze the Spiders, who keep their bodies limber by making their own antifreeze.

Like Wolves, Spiders are apex predators—they're at the top of the Insect food chain. Some Spiders even hunt like Wolves, by stalking up close and then dashing in for the kill, which is a classic example of parallel evolution. As with Wolves, I've learned stalking techniques from Spiders that I have used on Grouse, Bears, and humans. I'd like to tell this story in honor of Spider, who along with Wolf has been my esteemed tracking teacher. I have been blessed to live with both of them, and by shadowing them I've been able to walk in their footsteps and share in their adventures. They have done much to awaken the intuitive tracker within me, and at the same time they've inspired me to incorporate more of the tracker's awareness and spontaneity in my daily life.

It all started one winter afternoon in my youth when I was sitting at the window watching a flock of Cedar Waxwings feeding on Hawthorne fruit. However, they weren't just feeding. There was a social interaction going on and I was trying to figure out what it was about. That is, until I became mesmerized by a little Wolf Spider stalking a Fly on the windowsill. And I mean little—that Spider was going to have his hands full if he got a hold of the Fly.

Without trying, and without even realizing it, I became Spider. My chest tightened as I felt the dynamic tension he kept so well disguised by his outwardly relaxed state. I adopted his keenness of focus and at the same time maintained overall perspective. Every cue, every minute movement and sound and feeling were picked up. Together we synthesized the information and moved accordingly.

The more I became Spider, the clearer I could see that Spider had also become Fly. Again without effort, I felt myself becoming Fly as well. However, it didn't stop there. I realized Spider had also become the dust and dapples of sunlight on the windowsill, the cold draft seeping through the crack under the window, and the shadows of the fluttering Birds. We were all not only in this drama together, we *were* the drama. We were in the most intimate of relationships—the dance of life and death.

I didn't dare blink. My senses were keened to every movement, whether it be the smallest flutter of a dry leaf on the branch outside the window, the Fly changing her posture ever so slightly, or the ripple of disturbance created by the appearance of another Fly. I was prepared for anything, from an agonizingly slow stalk to pouncing as fast as a sprung trap if the Fly spooked.

Not quite as cool and centered as my mentor, the Master Stalker, I broke into a nervous sweat. My eyes felt dry and fatigued, and I worried that my movements were growing less and less fluid the closer I approached. Would I pounce too soon, or would I miss because I was too tense?

But that was Tamarack. I had to let him go—I was Spider. So imperceptibly, so slowly, and so in sync with the greater movement, I crept closer. And closer. There were times when I might have moved the tiniest bit, but I wasn't sure. I was a magician duping the audience into thinking no sleight-of-hand occurred. And still you knew something happened, because after all I was a magician.

I made it to about three [spider] body lengths from my quarry and she took off. Was it me, or did she have some reason of her own to bolt? I didn't know, and I didn't give it any thought. No regret or self-blame—I was hungry! Immediately I settled back into cultivating the illusion of benign presence and waited for the next Fly.

Benign presence does not mean invisibility. It means being so nondescript, so much a part of the landscape and mood that one can be looked at and not seen. Invisibility—especially when aided by camouflage—is usually easier because it is passive. I'm relying on something other than myself to help me disappear. Benign presence takes active involvement. It requires an unwavering commitment to leaving my sense of self behind and becoming my surroundings. The stakes are high, because I'm exposed and vulnerable. In this case, I'm out on a wide-open windowsill and don't have the option of invisibility because I need to get close enough to pounce.

A pack of Wolves might chase ten Moose before bringing one down. Wolf and Wolf Spider have the same hunting spirit. For them, a miss is not a failure because they live not only to eat, but to hunt. If they could live just by grabbing the first hunk of meat to come along, they'd grow dull and weak. Since they have to constantly work for their food, they're given continual opportunities to hone their hunting skills, and this keeps them growing stronger and brighter.

We modern people seem to have forgotten that this way of real-life learning and growing is also our way. We've accepted books and teachers and courses as substitutes for experiential learning. These methods may give the illusion of learning, but only because we live in a society that is built on illusion. Yet the hunter in us knows better. Rick, my assistant, just returned from eight days alone in the wilderness, living on only what he could hunt and gather. He told me how useless his well-developed deadfall and snaring skills were because he didn't learn them when he was tired, cold, and hungry. He now knows what no book could teach him—that it is hunger that makes the hunter.

The hunt works both ways: in exchange for the food that Moose and Fly give Wolf and Wolf Spider, they keep their providers healthy by weeding out the weak and slow. The feeble Wolf and Wolf Spider get weeded out as well. Only the most able survive to pass their traits on to the next generation—a vital contribution to the well-being of their kind.

When we live in our evolved niche in the Hoop of Life, we serve our prey in the same way as the two Wolves. In fact, they are our hunting companions. This is readily apparent with Dog, who is a domesticated Wolf, but not many of us are familiar with the role Spider plays.

Without Spider to help me track, there are times when I'd only be able to hobble along after an animal. Spider gave me speed when she taught me *web tracking,* and it came in handy one recent mid-afternoon when I was headed out to the wilderness camp where the students live. Usually I'll take what I call the teaching trail, which could be anything but the old Deer trail people usually follow into camp. I will continually challenge myself by taking unfamiliar routes, rather than relying on old knowledge and familiar habits. This way I keep learning and discovering. Still, I'll sometimes crisscross the Deer trail or even walk a portion of it, to pick up on any sign of unusual disturbance and check in on students' comings and goings.

I walked into camp and greeted Pam, a quiet, introspective student who had been here for about three months. Sitting over at the hearth, making cordage, she was the only one of the six in her camp to have returned so far from morning orienteering exercises and firewood gathering.

"I heard you got here at about midday," I said.

"How did you know?" she asked, assuming there was no one around to have let me know.

"Spider told me," I replied.

I could've talked more with Pam about my tracking her walk in, but I chose not to. I could have told her I saw where she stopped to admire a coral mushroom, and that she tried to catch something by the little pond, but she had more core skills to learn before Spider could help her see what was possible with web tracking. When the door is opened too wide too soon, people sometimes get overwhelmed and it can hinder their growth.

Spiders string webs just about anywhere they can be secured, and sometimes that seems like everywhere. If we were to look at this plethora of webs as a curtain, we could use tears in the curtain to show us when and where animals moved through. The breaks form a tunnel, which sometimes make it possible to track an animal by spiderwebs alone. The height and width of the breaks give the size of the animal, the type of break gives the direction, speed, and mood of the animal, and the age of the break gives the time the animal passed through.

The intuitive tracker will rely upon one type of sign only if she has to. Many of us like to use the rule of three, which is finding three distinct signs that are unrelated and say the same thing. The intuitive tracker will often use the rule of three without even thinking about it, as she just needs to barely glance at the sign. For her, tracking has become an art form. She can move swiftly after the animal because she is not burdened by having to stop, read, and compare sign. To the outsider, it could seem as though she just knows when the animal passed by, and where he is headed.

It is said that haste makes waste, but it's not so when it comes to intuition. Usually one's first impression is right on, and it's usually when we mull something over that we come up with other options and end up discounting our intuitive voice. How often have you kicked yourself because you didn't go with what first came to mind?

When it comes to intuitive tracking, the less we analyze, the more we learn. Along with exercising our intuitive abilities, we're challenging the deeper parts of our mind and learning how to use our ancestral memories. We're training our senses to pick up things quickly and our brain to process quickly.

How can we learn to read spiderwebs? By asking Spider to teach us. We can begin by watching her make her web. And then watch a hundred more being made, and another hundred, noting how each species makes her web and where. What is her preferred season? What is the time of day and weather? Is she with eggs or young? Has she eaten recently? Does she have a food cache? Is she building a new web or repairing an old one? Each of these factors plays a role in how fast she builds her web, along with the type and size. This was my training, and this type of information is what told me about Pam's walk into camp.

Again, web tracking can be learned best—and I'd venture to say only—from the Spider clan. But not all the clan. When we think of Spiders, we usually associate them with webs. The fact is, only about half of the Spider species are web spinners, yet all of them produce silk. It is an amazing material—the strongest natural fiber there is. It is said to be stronger than steel or Kevlar and more elastic than nylon. Silk can stretch up to one-third its length without breaking—which anyone knows—when walking into a web.

One thing you might not know is that a part of a web might be sticky, and another part might not be. This is because some Spiders produce sticky and non-sticky silk. They will use the sticky stuff to make the inside, or working part, of their web, and they will make the outside, where they hang out, with non-sticky silk. This keeps them from getting trapped in their own webs, yet it sometimes happens. Fortunately, they can cut themselves out with special sharp claws on their legs.

This strength and stickiness is important to the tracker, especially when spiderwebs grow old. Commonly known as cobwebs, these Elders have seen a lot pass their way. We young ones sometimes forget that knowledge can be picked up easily, but wisdom comes only with time. Rather than ignoring them in some dark and dusty corner or sweeping them away, it would serve us well to sit respectfully before them and listen to their stories.

The students attending a week-long outdoor skills course that I ran a number of years ago heard a story from a cobweb that I'd like to share with you. We were out on a hike focusing on one of my favorite things: nothing. I asked the participants to point out whatever caught their fancies, whether it be animal sign, a plant whose edible or medicinal properties they were curious about, landscape features . . . anything was fair game.

After checking in on what a Bear had been eating by picking through a nice fresh pile of his scat, which we found at the edge of the bog, we climbed a small rise and poked around in a grove of Hemlock Trees. A couple participants came across a small den under the roots of an old Tree, and a couple others found a similar den under another Tree about five paces from the first.

"No one has used this hole in a while," someone commented. "Look at these cobwebs over the opening."

"Same with this one," added a person by the other den.

A cobweb is a spiderweb with history. Sheltered from the weather, a cobweb can last a long, long time. It protects itself from being broken down by bacteria and fungi by being highly acidic. Over its lifetime, it acts like a historical archive, collecting artifacts that tell the story of the changing seasons and events that occurred.

"Look at the leaves on this web," someone pointed out.

"This looks like a Dandelion seed," suggested another person. "No, it's bigger than that, maybe it's Milkweed."

Someone else pointed out a feather on the web.

"Tamarack, can you help us figure out what these things are telling us, if anything?"

Typically I would play a guiding role in their exploration, being more of a catalyst than a teacher. However, this was a one-week intensive, and without having a working relationship with them, I decided I could serve best by taking on the conventional instructor role.

"How about backing up to gain some perspective first?" I suggested. "Get a feel for the area around the den, and then look at the web without focusing on anything *in* the web. Give it a couple of minutes and then let us know what you heard."

Old age was the most common impression. Compared with the surroundings, this protected nook seemed to have sat undisturbed for a long time. But not entirely. Someone noticed a small hole at the base of the web, off to one side. It was three fingers wide, about the size of a Chipmunk, somebody else thought.

We took a closer look at the sign. With the help of some guiding questions, the group interpreted the clues to mean the den had probably not been entered since last summer or early autumn. Milkweed had not yet gone to seed, so the seed on the web had to be from last year or before. After learning how to tell the difference between a withered summer leaf and a naturally shed autumn leaf, they figured the leaf below the Milkweed seed was most likely from last autumn. The thin layer of leaves that carpeted the den entrance was leaning up against the web, and they were autumn leaves, so the web had been constructed before the Trees shed. They thought the feather could have come from either this year or last year, because a birdwatcher in the group knew we were in the molting season.

"There's a story within a story here with this feather," I suggested. "A Bird has thousands of feathers, yet any one feather can tell us a great deal about the Bird it came from."

The group grew silent and focused on the feather I had just lifted off of the web.

"Feather reading," I continued, "is the tracking skill I use to expose the Bird's story. In another class someone found a Woodpecker feather and we talked for an hour about the Bird's life and death—all from what the feather told us. And there was more; we could have kept going. To get a real feel for feather reading, we'd have to devote a whole workshop to it. It's not that it's that hard to learn, there's just a lot to it.

"I'll give you an example if you'd like to hand me that feather on the web. It's called a covert; it's one of the small feathers that give contour to the wing. Here, pass it around so you can see what I'm describing. The shape and size say it's from the left wing, about three-quarters of the way up. She dropped it right around halfway through her wing molt. Unfortunately, we don't have other coverts to compare this one with, for if we did I could easily show you how this feather told me what I just told you. However, you might be able to see why this feather is from a young Bird rather than an adult. Notice how dull this feather is. If it came from an adult, it would have more sheen."

"So what kind of Bird is it from?" someone asked.

"If I told you, you'd say, 'Oh,' and that would be that. If you did some research and found out on your own, you'd probably learn a lot more about the Bird than just the name. And you could experience what I consider to be one of the great benefits of tracking—the joy of discovery. Let me give you a clue to get started: the color and pattern of this feather are the same on the whole shoulder patch. Keep in mind that research is tracking, because you'll be using some of the same skills we use here in the bush.

"To keep your research relevant," I continued, "I'd suggest staying as close to the actual Birds and their habitat as possible. When live Birds are hard to come by, you can check with nature centers, universities, and museums that have mounted specimens and skin collections. As a last resort there are books, which in cases like this, I think are okay to use. They are augmenting, rather than substituting for, direct experience."

I wanted to show them one more thing to give them a start, so I pointed out the three indentations that ran across the feather about half way up. The grooves were so pronounced that they even indented the shaft. I passed the feather around so they could all take a look. The grooves, I told them, were caused by a protein deficiency, probably from missed meals, because the feather looked healthy otherwise. A poorly formed feather could indicate either a parasitic infection or a glandular dysfunction, where a feather defaced with numerous indentations usually means the Bird was the runt of the clutch. I added that an indentation half way up the feather shows the little guy missed those meals when the feather was half grown out.

I explained that a young Bird needs a tremendous amount of protein—and on a regular basis—since feathers are made of protein and they're all growing out at once. On top of that, muscles, skin, organs, beaks, claws—virtually everything other than bones—are made of protein. It's so important to a nestling that many vegetarian Birds who typically live on seeds and fruit will feed their young high-protein animal diets so they can develop properly.

Because of time constraints, I didn't get into the variety of storytelling marks a feather can carry other than protein-deficit indentations. A few different marks can tell us whether the Bird lost the feather because of the molt or because she was killed. A mark can even reveal who the killer was. Other marks can tell us the Bird's health, her level of physical activity, her favorite perching place, and more. And then there is the feather's color, pattern, and structure, which talk about her size, sex, age, marital status, and region of origin. I hoped that some of these workshop participants would be coming back for a course so we'd be able to explore these more subtle voices of the feather.

What I did do was ask them, "Why would a nestling miss a meal, especially three in a row?"

They came up with several ideas—food was scarce, the parents were gone somewhere (maybe a predator scared them off), it was stormy and they couldn't go out foraging. With two or three grooves tight on each other, I've found the last suggestion to be the most usual cause. I've checked young in the nest to see exactly when the indications form. Mama will sometimes sit tight on her little ones and not leave them when the weather is rough. This could end up being a couple days with a slow-moving low pressure system.

This feather speaks in another way, by how it lays on the cobweb. A spiderweb's stickiness lasts only so long, as it degrades with age and the accumulation of dust and debris. When we took the feather from the cobweb to look at it, we could have noted whether it was adhered to the web itself or whether it was caught by dust. Knowing this would have helped us determine when the feather came to rest on the web. "Instead, I chose to focus directly on the feather," I told the students. "However, there's no shortage of cobwebs in this world. You'll find plenty of them to continue learning from after this class."

The cobweb over the second den, they noticed, was only about half as dusty as the first, and there wasn't any debris on it. The excavated dirt looked fresher than that in front of the first den. They figured this den was occupied over the past winter and the cobweb had been woven over the den entrance since then.

So concluded our cobweb adventure. I was satisfied that the seed of inquisitiveness had been planted. I knew that when the students came across cobwebs in the future, there was a good chance they'd see the old bug snares for the storytellers they are. Those who continued listening would become more and more able to hear the song of the cobweb's track and read the sign at a glance. My greatest hope was that one or two of them would be inspired to follow our ancestral way and take Spider as their teacher rather than turning to a human or a book.

And the same with us—even if we're not able to take to the wilderness and hunt with Wolf, we are not deprived. We can learn the same skills in the wilds of our backyard or windowsill. There is sure to be a feather or spiderweb waiting to tell its story. And there is bound to be a Wolf waiting to guide us.

7
A Winter Riddle

Along a Frozen Lake, a Kill Leaves More Questions than Answers

Al, an energetic student in our wilderness program, recently found a kill site. Actually, he just stumbled upon it, and he couldn't make heads or tails out of it. So he went back to camp to get his tracking buddy and fellow student Justin, to see what they could figure out together. They make a good discovery team, with Al's spontaneity and straightforwardness and Justin's studied, reflective approach.

They say two heads are better than one, and sure enough, they both came to the conclusion that . . . ah, they couldn't make heads or tails out of it. When I talked with them last night, they asked if I'd come with them this morning to take a look.

It's the middle of the white season and the lakes are windblown and crusty, which makes it easy to get around. Earlier in the season, we got a lot of snow and it lay deep out in the open, so we had to break trail wherever we went. Now we can head out across the lake in front of camp just about as easy as walking through a park. Only we don't have to watch out for Dog poop or Frisbees.

About fifteen paces out on the lake, we come across Coyote tracks in the light snow covering the ice. They parallel the east shoreline, but we had wanted to cut over to the kill site on the northwest side. The lure of the trail prevails—we follow the Coyote tracks.

"I bet he's walking out here for the same reason we are," says Al. When the first light snows came, he and Justin would pick up tracks of one or two Coyotes working the wooded shoreline. Now the snow lies deep under the Trees, so they've taken to following the shoreline on the ice.

"Well crack my ass," says Al with his typical flair as he gets down on his hands and knees to smell the yellow stain on a protruding chunk of ice. "Who'da guessed he'd be marking territory out here? What does he do in the summertime—dogpaddle out here, lift a leg, and soak a wave?"

The way these boys joke around, you'd never guess one of them was a submarine physicist and the other a Special Forces operative and

Zen monk. Actually, it is probably good they threw their professional demeanor out the window, because it would only get in the way of their intuitive guidance and ancestral memories.

With Al down there smelling pee, you might be thinking he threw a little too much out the window—especially if I told you he has also been known to taste it. He is actually practicing urinology: the study of urine for clues to diet and health, which is a legitimate scientific discipline. The accomplished tracker can read even more with scatology.

Yes, in our analysis we get a little more *intimate,* you might say, than the average lab technician would, but look at all the time and money our on-the-spot analysis saves over sending specimens to the lab. Through urine smelling and tasting, our students learn how to identify species, gender, stage of breeding, diet, and sometimes even health and activity level.

I'm sure this sounds less than appetizing to some of you, but fresh urine from a healthy animal is sterile and safe. Many of our animal Relations analyze urine. In fact, they're the ones who taught me. My Wolf kin were particularly good at it. Only I got lazy when I was with them, because they'd just go ahead and do the testing and give me the results.

We all take a sniff of the Coyote pee. In order to pick up its essence, we have to breathe on it—a technique I tell the guys is an old Indian trick. My students used to roll their eyes when I'd use that line, and now they just ignore me. But I don't mind; I'm still having fun.

Is it an old Indian trick? Beats me, yet I'd bet my life on its effectiveness. I learned it from Wolf and I know from Elders' stories and research that some young Indians learned to hunt from Wolf, so I'd be surprised if a few of them didn't pick up scent releasing just like I did.

It works because air that is cold and low in humidity isn't a good scent carrier, and the colder an object, the less scent it releases. Breathing on the object brings its odors to life by warming it and adding moisture to the air.

We cut across to the north side of the lake, and just out from the shoreline, we come across more Coyote tracks.

"Do you think this is the same animal?" I ask.

"Are you kidding?" responds Al, the quick one. "Look how old these things look."

"Watch out," says Justin. "I'd say you just stepped into a trap."

I smile.

We go over what they learned so far this white season about how tracks on the lake are affected by wind, sun, and water seeping up through cracks in the ice.

"Is that what aged these tracks?" I ask.

"Another trick question?" asks Al.

"Not at all," I reply. "All these things were factors, yet there's something here that magnified their effect that is not on the other side of the lake."

They look around to see what is different about the two sides of the lake, and they don't come up with anything.

"Remember where you found the warmest water early last spring when you wanted to go swimming?"

"Duh!" says Justin, smacking his forehead. "This side gets no wind and lots of sun. I've been over here a couple times lately, on sunny afternoons, because it felt so nice—almost like spring. I even had my coat off. Yeah, I can see why these tracks would look older than they actually are."

We follow the Beaver canal that cuts back from the lake into the bog. To the left of the canal is a steep bank that rises up to a plateau forested with Maples. To the right is a typical Northwoods peat bog peppered with stunted Spruce. Justin says the kill site we are going to see is just to the right of the canal, about forty paces in from the lake.

What I envision from the guys' description the night before is a patch of scattered, trampled feathers. Does this mean it is a kill site? They think so, and it could well be. On the other hand, a bunch of feathers on the ground is actually one of the least reliable signs of a kill site. The Bird may have been attacked and escaped, she might have been carried there from elsewhere, or she could have been already dead when the predator found her.

So what makes a kill site? A scuffle? I've seen predators leaving "scuffle" tracks by tossing dead animals around and playing tug-of-war with them. How about the tracks of both predator and prey entering the site and only the track of the predator leaving? Again, the prey animal could have died of other causes. And then there's the "kill site" my brother Bill came across, where only the tracks of the prey animal left the site. He watched a Weasel jump a snowshoe Hare and latch onto her neck, and the Hare took off and gave that Weasel the ride of his life!

Does this mean there's no reliable sign to confirm a kill site? If there is one dependable sign, I either haven't yet discovered it or I'm not a good enough tracker to trust in it. And then again, maybe I'm just too good a tracker and don't know when to stop questioning.

Rather than sign, I usually go by feel. It is something like picking up on the presence of a park in your new neighborhood before you know about it because of the sounds of children playing and the joggers and Dog walkers heading in that direction. Odds are you don't think about it—you just know there is a park over in that direction. You might say you're picking up on the song of the park's track.

As we approach the alleged kill site, Justin tells us that yesterday he completely circled it at about fifteen paces out. He did this in order to survey all of the animal trails coming in and out of the site and not disturb any of the sign close to the kill. He says he was hoping to identify the various scavengers coming in to feast on the leftovers and maybe—if luck was with him—ID the killer himself. "I'm not sure what happened here, Tamarack," he concludes. "I spiraled in sun-wise the way you showed us, and I came across lots of tracks. There was Fox, some big Weasel-looking tracks, and even Squirrel. But I didn't come up with a friggin' clue to how the Bird was killed or who killed her."

We come across Justin's first round of tracks and I feel myself shape-shifting into a predator. I'm no animal in particular; I've just assumed the consciousness common to all hunters. My senses are keened, I gain perspective on the area, and I listen intently to my intuitive impulses and the song of the place.

When we come within sight of the kill, we find a place to stand where we won't be stepping on any obvious sign. Justin and Al point out what they found the day before: chewed feathers, a small faded-out bloodstain, and the various tracks Justin told me about. Next they explain how they aged the site, based on the depth of snow over the oldest tracks and when they remember the snow falling.

I pay close attention to the guys' presentation, and at the same time I feel a tug from up on the bank behind us. The lingering notes of a song—a lyrical ballad of life and death—come clearer the longer I listen.

When they finish their story, I turn around and climb about three body lengths up the bank and motion for Al and Justin to join me. "Did you check up here yesterday?" I ask.

"No, I didn't think of that," replies Justin, and Al nods in agreement.

"Why don't you sit here for a bit," I suggest, "and see what speaks to you?"

We are underneath the low branches of a stand of young White Pine and Balsam Fir so dense that the needle-carpeted ground is bare of snow. It is another world, completely invisible from above and unknown to anyone walking by unless they crawl under the protective branches around the perimeter of the stand like we did.

Diverting my attention from the obvious sign, I sit there noncha-lantly and listen to the song of the track while they work their eyes over the groundcover. As laid-back as I might look, it is only a smoke-screen—I'm listening with intense interest. And I'm listening with-out prejudice, as though every voice were a precious gift, a vital clue. Respect is the cornerstone of hearing the song of the track. Without regarding all that I hear—whether or not I like or agree with it—I'll miss important information. Those with selective hearing, those who filter out or discount anything, sooner or later meet their demise. Sure, they might survive for a while, but they will not thrive.

The consummate tracker is like a sponge that absorbs everything and anything within range. When I struggle to hear the song of the track, I know it is not for lack of ability as much as it is for lack of respect. When I can step out of my ego, I begin to hear what I once filtered out because I judged it unimportant or unfamiliar.

My listening is interrupted by the guys' intense focus on several fist-sized mounds of scat they come across on the ground a few paces away.

"This looks like Bird scat," says Justin. "Is it Grouse?"

I only smile.

Neatly deposited atop one of the mounds is a dark, rich-looking scat that looks like a swirl of chocolate topping.

"How sweet," says Justin mockingly. "It looks so good that Dairy Queen could run it for a special. Let's see . . . what could they call it? How about Double-Butt Colon Crunch?"

Ever the jokester, that boy.

Anyway, the two have learned enough to know that the scat's slick, chocolaty appearance means it is from an animal with a high-protein diet. And they know that the scat's elongated shape and stringy-ended cone say, "Weasel family." However, they haven't yet learned enough to know which one of the family it might be. Nor have they seen enough to realize it is not uncommon for members of the Weasel clan to take dumps near—or even on—someone else's doo-doo. To me this usually appears to be in response to territorial threats.

For a tracker, scatology isn't just about the scat itself, but also the accompanying sign, such as location, the direction the animal faced when squatting, digs and chews, along with sign left by other animals attracted to the scat. One little pile of poop can tell quite a story! The ability to hear it can be especially valuable for the beginning tracker, as scat is often the most visible and easy-to-find sign.

To clue you in on what Al and Justin are seeing, let me give you some of the scat anatomy basics that they learned while they were here. There are three parts to a scat pile: the plug, the body, and the cone. The firmer the scat, the easier these parts are to distinguish and read, and the softer the scat, the more difficult they are to read.

The cone, which I just mentioned regarding Weasels, is the tail end of the scat. It usually tapers down to a point because it isn't ejected with as much force as the rest, and because it gets pinched off. You'll see this exaggerated with the Weasel family, which is why their cones are long and stringy.

The shape of the cone and how it lies in relation to the rest of the scat can tell much about an animal's emotional state, such as whether she was nervous or relaxed, and if she was taking her time or in a hurry. The cone shows the direction she faced when she squatted, which could say whether she was confident or fearful, and even what she feared. She was in a vulnerable position, so she may have chosen both a location and a direction to face where she could remain vigilant—she was likely in a state of hyperawareness.

The plug is the first part of the scat to come out. It is usually the driest part, because some of the moisture has been reabsorbed by the body. This makes it firm and compact, so it often breaks off. It can stand out because it is more fibrous and dull looking than the rest of the scat. The head of the plug is called the bulb, and it is rounded and bigger in diameter than the trailing end. However, Weasel-family bulbs can be tapered, much like their cones but not as extreme.

Between the plug and the cone is the body, which makes up the bulk of the scat. It is usually softer than the plug and a little smaller in diameter. When taking scat diameter measurements, use the body, as it will give you the scat's overall average width.

If you want to find out what the animal has been eating recently, dissect the body of the scat. Even though breaking apart the plug is usually easier because of its dryness, it contains only remains of the oldest food eaten.

"So what do these scoops of maple-nut ice cream mean?" asks Justin, more to himself than to anyone else. "I'm not hearing anything."

As his voice trails off, something draws him to look up. "Ah, a perch!"

This is the breakthrough they need, but at the same time I'm concerned that too much knowledge could backfire. Now there is a good chance that they'll quickly discover more clues, get excited, and jump to conclusions. This could set them back in their evolution as steady, intuitive, questioning trackers. I suggest that we ramble on, and that they come back on their own later in the day to see what else might speak to them. They might then have the perspective to view the area not as a kill site, but as a site at which many things happened, including perhaps a kill. Rather than just keying in on a single voice, they might then be better attuned to hearing the whole story.

Imagine picking up a novel and going through the whole thing but reading just one character's lines. This is what we do when we focus on just what we're able to see.

When Al and Justin return, they'll probably find the four small, iridescent neck feathers that were torn from the Bird when she was pounced upon. I bet they'll also notice the disheveled needles from the very brief scuffle that occurred when they both fell out of the Tree. I'm hoping this'll give them a feel for the continuum, as opposed to creating a stop-action event of the kill. This is what they've been learning to do over the eight moons they've been here in camp. If they succeed, they might be able to become the Fisher who killed the Grouse and feel what he felt. There is the hunger that drove him, the intense, immediate desire to satisfy that hunger, the adrenal rush leading up to the kill, and then the warm satisfaction of the feast.

The most challenging part of their envisioning is probably going to be recognizing that the killer was a Fisher. Members of the Weasel family are sexually dimorphic, which means that males and females are of different sizes, and in this case, males are bigger than females.

From smallest to largest, the land-dwelling members of the family in our area are: Least Weasel, Shorttail Weasel, Longtail Weasel, Mink, Marten, and Fisher. A male Longtail Weasel could be the same size as a female Mink, and a male Mink could be the same size—or even larger—than a female Marten, and so on. To add to the confusion, the tracks and gaits of these animals are similar enough to each other that the novice tracker could easily misidentify them.

In these cases, I'll consider the robustness of the paw print. A male Marten's print will be more delicate than a female Fisher's of the same size. Next I'll look at secondary sign, such as habitat. Meadow and marshland says "Least Weasel," while shoreline would probably be Mink, and mature forest is usually Marten.

After the kill, the Fisher would immediately look for a safe place to feast. Justin and Al, having slipped into Fisher consciousness, might just pick up on this. Right now the Fisher's worst fear would be having someone come and threaten to take his meal. If he tried to eat in the thicket, he'd be nervous, because he wouldn't be able to see if anyone was coming. There, just down the hill in the bog, he sees a little opening where he can eat in peace. It is shaded, which he likes, and still it gives good visibility.

If the guys pick up on what really happened here, I can just hear Al saying with his "Joizey" accent, "Ah, an' here we tought we had a kill site. Hah, it was jest scraps left on da dinnataeeble." I bet that'll be followed up by a resolve to *be as a question* right from the get-go from then on.

But that'll be later—right now we're ambling on. "I've got another banquet table to show you," says Justin, who leads us out across the bog to a dead Deer he came across a couple of days ago. Judging by how trampled the area is around the carcass, along with the number and variety of tracks, every carnivore in the area must be feasting on this kill. Even a few herbivores have given it a try.

This isn't so uncommon, actually. Many animals commonly considered to be vegetarian, such as Deer and Rabbits, will occasionally nibble on meat for the concentrated nourishment. Ground Squirrels will even catch and eat small Birds. The reverse is true as well, and Pine Marten is a good example. Known for his ability to run down lightning-fast Red Squirrels, it turns out he is quite fond of blueberries—so much so that his scat turns blue at the peak of berry season.

Coyotes have been coming in regularly to scavenge, and Justin wants to know if they're some of the ones he hears howling at night. He gets a bead on where the howling is coming from, and the next morning he goes out to find the site. He says he looks for clues that'll tell him if they howl when they get together for the hunt, or if it is after they make a kill, or if it is after they feast on the kill. "And then again," he says with a wink, "maybe it has nothing to do with hunting. You've got to be as a question, you know."

Justin points out one well-used Coyote trail that comes in from the east. It winds through a young forest of Aspen, Red Pine, and Balsam Fir at the edge of the bog and then tracks off into the wilderness. "Yesterday," he says, "I came across a Coyote's track several miles out in the wilderness and followed it. Wouldn't you know, it connected with this trail. Seems like Coyotes are coming in from all over the place. How do you think they all found out?"

"Howl if I know," I reply.

He groans.

"All right," he shoots back, "I got another one for ya. About a quarter mile up this trail I saw some tracks, but I couldn't figure out what they were. Wanna take a look?"

Neither he nor Al could tell at a glance who made them. Similar to the tracks we saw earlier on the lake, these were distorted by the sun. On top of that, they were worked on by the wind and grew with time. Yep, they grew—tracks in the snow are notorious for unbelievable growth spurts. The snow is to blame: it has this odd quirk of disappearing into thin air without first needing to melt. The stronger the wind and the drier the air, the faster it evaporates. Although the snow shrinks, the tracks imprinted in it, of course, grow.

I remember once finding a nice, clear Bobcat paw print out in the open. It was from the left front paw, and it was a big one. She looked to be headed for the Fir Tree right in front of me, so I ducked under the overhanging branches and looked around for some sign. All I could find was a Snowshoe Hare track. Boy, was I glad nobody was along with me, so I could keep that one a secret. Shh, don't tell anyone.

I stand back to give the boys plenty of space to study the tracks. They look for one that shows a heel or paw pad, or maybe claw marks—something distinct enough to help identify the animal. I don't bother to look. On the way into the area, I gained a feel for the animal's sex and age, and it was obvious at a glance as to what he was up to, where he came from, and where he was going.

"I think it's . . . maybe . . . a Coyote track," Justin offers, even though they didn't find anything clear enough to go on.

"What do you make of this animal meandering back and forth from this Hazelnut bush over to that clump of Grass sticking out of the snow, and then to the next one?" I ask, hoping they'll pick up on the hint to step back and gain perspective.

"Now that you mention it," says Al, "this thing ain't acting at all like a Coyote bee-lining it for a steak sandwich."

"Maybe he got sidetracked," Justin throws in.

"Good point," I reply. "Anything's possible."

"You sly Dog, Tamarack," says Al. "You know what's going on here! Why don't you quit playing with our heads and just tell us?"

"And spoil all my fun . . . er, I mean, your fun?"

"All right, give us another clue."

"Okay, but don't groan—you asked for it. Be as a question. Forget Coyote—forget everything—and become the animal who made this track. If you slipped into this animal's consciousness, you'd see that this place is dripping with clues."

They listen, and I can see them letting go. They want to learn, or I should say *unlearn*. As before, I step back to create space for their process.

Justin and Al end up getting stuck—they need something to break the impasse, something to get them out of the blind tunnel their brains have taken them into. I point to a footprint in front of them that is unique because it has broken the surface. The animal stepped on what appears to be a mound of Moss, which in actuality is a Moss-covered rotten stump. The animal's foot broke through the Moss and sunk into the punky wood.

They look at it—they get close, they back up, but nothing clicks.

"Don't think," I say softly. "Be."

"Ah!" says Al after a bit. "It's not a paw print—it's a hoof print. Correction: it *might* be a hoof print. A paw could be supported by the Moss, but a hoof would cut through."

This opens the door and they start putting the pieces together, even those they didn't notice before.

"So that's why the 'Coyote' wasn't bee-lining straight for the kill," states Justin.

"And look at this Grass," adds Al. "Our 'Coyote' wasn't just sniffing around; he was chomping the stuff down!"

This story had a happy ending, and lessons were learned. If it were always the case, I'd be a happy puppy. The reality, however, is that most of the students I work with get stuck. And if they don't have guidance, they usually stay stuck. Why is this, and why is it important for them to get beyond this point?

Awareness is the first and most essential step to change. If I don't know where I've come from or where I'm going, I'll likely end up walking in circles. On the other hand, when I know where I've been and where I want to be, I can set a direction.

My experience with new trackers like Al and Justin shows that their biggest stumbling block is the urge to push right away for an answer.

Most of us start establishing this approach shortly after birth, when we're isolated from daily goings-on in nurseries and cribs. Starved for stimulation and real-life involvement, we dive head-on into things at every opportunity. Once the pattern is set, we continue with it after we leave the crib and usually carry it with us throughout our lives.

A native child's first turn of the seasons is commonly spent in a cradleboard or sling, where he is continually with his mother and in the center of activity. From this vantage point, he observes and gains perspective, which becomes a lifelong practice. This is the natural way of child-rearing for humans and other primates.

One reason our dive-into-it pattern is hard for so many of us to kick is that we keep fueling it with stimulants. Coffee, soda, sweets, music . . . the list goes on. These pick-me-ups keep us in our egos, keep us revved up. They make it hard for us to slow down and listen to our intuitive voices, or to hear the subtle song of the track. We are living on the surface, using just the outermost layer of our deep, deep minds and only the very edge of our sweeping senses. With most of our capacity shut down, there is not much else we can do but get analytical and grope for an answer. We're simply too numbed out and disconnected from ourselves and the Hoop of Life to notice the subtle voices that have so much to tell us.

After we quit stimulants, as have Al and Justin, the habits we developed during our stimulant-using days can stick with us for a while. This is why Al and Justin keep hitting walls. They continue trying to get their frontal lobes to do the work of their senses, intuition, and ancestral memories. No matter how clever and knowledgeable they are—and they were both successful in their old lives—their minds alone cannot make them fully alive and functional humans.

Relying on our rational processes is what we typically do when returning to the natural world. It is to be expected—we were born into it, we were trained for it all our lives, and it is how we have always gotten along. On top of that, the modern world is an artificial world, dreamed up and built by rationally dominated minds, so of course people living in that world need to rely mainly on their rational minds to survive in it.

At the same time you see Justin and Al struggling with this, you see signs that their natural selves are breaking through. They grow "curiouser and curiouser," and more and more they take the time to listen and make the effort to step outside of their "Justin" and "Al" identities so they can listen. It is both a pleasure and a privilege for me to work with them and their fellow students and watch them blossom into their original selves.

8
Eyes That Shine
The Tracker Can Now Shed Old Skin to
See with New Eyes

Have you ever noticed how some animals are so beautifully camou-flaged, with the color and pattern of their fur, feathers, or scales helping them practically disappear, and then—of all things—they have brightly colored eyes? To top it off, some of them will have stark white or black, or even fluorescent green, stripes right across their faces, with their eyes smack-dab in the middle. What a contradiction, don't you think?

Yet there must be a reason, as everything in the natural realm has a purpose. Let's go back to a brisk, sunny afternoon in that dynamic tran-sition time between green and white seasons. I'm paddling across the lake to check in with our second students' camp. Before I have my boat pulled up on shore, I see Antoine coming down the bank toward me. His focused eyes and deliberate steps tell me this is no regular greeting.

Antoine, who has been a student for about half a turn of the sea-sons, came here straight out of high school. Idealistic and dedicated, he dove right into everything the program had to offer. He was inspired by his older brother, who was a prior two-year student. Their ancestry is Canadian Ojibwe, and both came here to reconnect with their hunter-gatherer roots.

"Let's go sit up by the fire," I suggest, "and you can tell me your story."

"It happened this morning," Antoine begins. "I was cutting across this little bog out in front of us, and there was a Deer in that little stand of shrubby Bog Birch right over there where the highland comes down to meet the bog. I could barely make her out," he continues. "The color of her new winter coat, and even the texture, blended in with the bark and twigs. And she stood perfectly still. I couldn't make out her legs; their outlines must've merged with the standing brush.

"I could see just a little patch of her belly—or what I thought was her belly. I couldn't tell really, because I wasn't sure where her belly ended and the background began. I don't know if it was the color of her fur or the way it hung, but it just seemed to flow right into the dead leaves and sticks. The same with her back. Again, I couldn't see much of it, but the way it absorbed the light, it merged with the shadows.

"And then there was her eye," Antoine goes on. "I didn't notice it at first, but when her eye caught mine, hers stood out as though it was anti-camouflage—it was like a beacon drawing my attention. It didn't make sense—there was a perfectly camouflaged body, and then an eye that stood out like a glistening chocolate-brown marble. If I didn't watch her walking into the brush, I'd never have known she was there. What's the deal? Why would she advertise her presence?"

"Is that what she was doing?" I ask.

"It seemed that way."

"To whom?"

"Well, to me."

"Your question has an answer, but you're asking the wrong person. If you want to know what it is like to paddle a canoe, what would you do?"

"I'd go paddle a canoe. Ah, okay—to know what the Deer's up to, I could become the Deer. If I came down the bank to the edge of the bog like she did this morning, maybe I'd find out what it feels like to be on the verge of stepping out into the open. I'll just go and do it. But there's still that eye. . . . It doesn't make sense."

"To whom?" I ask again.

"To me, of course."

"Yes, and this is why you must become the Deer."

The next morning at about the same time, he strips off his clothes as I suggested, so that he will leave the secure trappings of civilized humanity behind and feel naked and vulnerable like the Deer.

Everything is related: our physical trappings tend to keep us locked in the headset that created the trappings. They become a shell, a comfortable protection. The trouble is that with protection comes isolation. The simple action of shedding our trappings, whatever they might be, can help free our minds, senses, and feelings from their conventional ways of functioning.

Later that evening, frost is in the air, and Antoine and I get comfortable in front of a warm and crackling fire. Here's the story he shares with me.

"While I was working my way down the ridge to the bog, I felt pretty comfortable and secure. The Trees were my companions—they gave me some coverage and protection, and at the same time I could see through them for some distance. However, that changed when I reached the edge of the bog.

"It was bright and open out there, and I felt apprehensive. Underfoot it was damp and cold, which heightened the feeling of change. I was making my way through brush so dense that I couldn't see around me like I could in the Trees, which made me all the more edgy."

The Deer, being a creature of the night, was probably even more edgy than Antoine. Deer's day sight is only average compared with other large animals, which makes them feel vulnerable and nervous. They much prefer being out at night, when they have sharp vision, along with superior hearing in the quiet. They have a keen sense of smell as well, and the still, damp night air is ideal for carrying scent. Sight is another story—they can't see details as well as we can. They move their heads to help make things out, which shows off their eyes and helps make up for their mediocre daytime eyesight.

"I waited there a bit," continues Antoine, "until I was pretty sure there was no one around, no danger lurking. Still, I didn't want to step out onto the bog. I had this urge to turn around and go back up into the woods. But something tugged at me—I knew there was a reason I needed to cross the bog. I stuck my head out and went one slow step at a time. My ears were perked, my nose was sniffing the air, my legs were ready to bolt me out of there in a flash if I picked up on anything threatening.

"Just ahead was a clump of stunted Birch. I wanted to run for them so I could hide there for a bit and get a break from the stress that was built up inside me, but I knew I had to keep going slow and steady. Being hidden in the brush would not do me any good if I made a racket getting there."

"How did it feel?" I ask.

"Tamarack—I was the Deer! It was cool—I was never so tuned in to the woods. And it wasn't because I wanted to be; it felt like I had to be. It was so much not like the usual, where I could just poke around and either be present or not. This was serious—somehow I knew that every move I made could mean either life or death."

He pauses for a few breaths, as though he needs time for the awareness to sink in. I ask him how it felt to move like a Deer.

"When I slipped into the brush," he replies, "I stood still—not a perfect, artificial still—but I swayed just a little bit with the branches and the spots of sunlight moving around. I thought I'd relax right away, but I didn't. I had new things to worry about. Was I alone on this island? What could I do to hear above the clattering branches and rustling leaves?"

Right there he breaks out of storytelling mode and looks hard at me. "Okay, Tamarack, this is where I got stuck. You say that in the natural realm everything is for a reason. Well, there I was, still feeling vulnerable even though I was under cover. How in the world could eye-catching eyeballs have helped me?"

"It appears you lost it, lost being the question," I reply. "It's because you lost your identity and became a human again—a thinking human, of all things. Now what can you do to get unstuck?"

"It's not that I don't know what I need to do," he says with a smirk. "I was just hoping to weasel an answer out of you. I'm actually looking forward to doing it again—it was a blast."

A couple of evenings later I am back at Antoine's camp and he greets me with a blurted-out, "It was amazing, Tamarack!"

We sit down at the fire and he grows serious before he speaks. "I couldn't see very far," he begins. "And there was a breeze again, so I couldn't hear much either. I felt vulnerable—every odd movement or sound caught my attention. I imagined the worst-case scenario, that there was a Wolf cutting across the bog, or a human. All of a sudden I didn't want to be camouflaged any more—I needed to know who was out there."

"How were you going to find that out," I ask, "*and* take care of yourself?"

"I wanted to be seen, and I wanted to see at the same time, so I could tell if that animal had any interest in me. If he did, I'd have time to escape. And if he didn't, I could sit tight and not draw any more attention to myself.

"And then a light went on—I'd want that predator to look me right in the eye. I could then have a bead on him just like he had a bead on me. And to grab his attention, I'd want my eye to shine."

All is quiet but the fire's hiss and crackle. Antoine seems lost in thought, and I gaze into the coals, glowing inwardly over the remarkable insight this hungry young tracker has gained. The shadows of the flames dance over his face, which makes me realize that this moment could just as well have been beside a campfire ten thousand years ago. Back then, we all lived by the rhythms of the Earth and the hungers of the heart. Adventures were shared, just as Antoine did tonight, and in this way teachings were passed down in story. I wonder if this story of Antoine's will be retold to young trackers gathered around future fires. And I wonder if their eyes will shine in the firelight like ours do now.

9
Grandfather Tip-Up

How to See Centuries of Movement in the Stillness of an Old Forest

The green season lingered this year. I watched a few Birch Trees drop their leaves early, but that was only because of the drought. Birch, being shallow rooted and moisture loving, is one of the first Trees to weaken when the rains don't come. This leaves them susceptible to disease and Insect invasion. Ah, but some Birch are wise—they'll shed their leaves to conserve energy and moisture rather than trying to beat the odds.

It is now the Falling-Leaves Moon and the Sugar Maples have finally gotten serious about contributing their lemon-yellow leaves to the forest floor carpet. I am still finding a few bunchberries to nibble on, which I don't remember ever doing this late in the season. Usually I'm out harvesting cranberries by now, but they still don't have the rosy cheeks that tell when they're ready. I tried a couple anyway, and they tasted as pale and acidic as they looked.

The season was off-track to begin with, as it got off to a late start and set many plants back: or so some would say. In the natural realm, there is no early or late, as no plants or animals keep track of dates or averages. The now is all there is, and whatever happens is just what is supposed to happen. We, on the other hand, like to have an "average season," where things happen in the same time and sequence as last year, the year before that, and so on.

However, this seldom happens. An average is not the norm; it is just a mathematical calculation. And if we had all average years, we'd be in trouble. Early and late seasons are crucial to the vitality of plant and animal populations. We can see this in plants that need a specific length of time to grow and reproduce. If they haven't gone to seed when the weather turns, they just keep going and hope for the best. Some of us see this as a race that a plant sometimes wins and sometimes loses.

This is another artificial construct of ours. Only in our competitive, black-and-white world is there winning and losing—with winning, of course, being the better of the two. To plants, losing is winning and dying is living. The challenge of the cold is welcomed because it makes them strong. Those who mature earlier or are more frost-resistant will survive to pass these resilient qualities on to their young. And the same is true of the survival challenges animals face.

These constructs just played their contrary magic on Rick, my assistant, recently back from a sojourn in Oregon, his home state. He marvels at how dynamic the change of seasons is here in comparison with the Northwest. "At home it gets wet, and then it gets dry," he says. "It gets a little warmer, and then it gets a little cooler. The leaves fall, and the leaves come back. Here when the Trees go bare, big things happen: the wind switches from south to north, the lakes freeze over, and you know you'll soon be wading through snow up to your thighs."

Rick was here for two Falling-Leaves Moons before he went back home, so he knows full well what to expect. It is not surprising that he'd be impressed by our seasonal changes, because they're so different from what he grew up with. All change holds beauty and mystery when we open up and immerse ourselves in it. If we didn't compare the new with the old, as Rick did, the magic of the new might escape us.

Since Rick has been back, he has been wandering the forest every chance he gets. He is studious and driven, with a passion to learn. Everything interests him, and there is hardly a time that I see him when he doesn't have a question about something he has come across. He has been guided by me long enough that he no longer expects answers, and he is now content with just a few clues here and there. At first he'd get frustrated—he is the typical ADD type who is shy on patience—but now he eats up the questioning approach. He has enough experience to know that he gains far more from someone's questions than their answers—like the time he and I were heading out to camp, and he was telling me about this fresh Badger den he found that was dug into the cut bank of an old logging road.

"Are you sure it's a Badger?" I ask.

"No, of course not," he replies. "But the opening was wide and low like a Badger's, and the sand was thrown out in a broad fan like a Badger does. Then again, it could've been two Woodchucks digging side by side and spreading the sand out wide to make it look to a hungry predator like an ornery Badger lived there."

I give an exaggerated look of serious consideration.

"But seriously," Rick continues, "I'm trying not to assume it's a Badger den and see everything through a Badger filter. I've done it before, and I'll probably do it again a time or two—or twenty."

We both chuckle. Rick says the people here should all wear T-shirts that say *Maybe* so I don't have to keep reminding them.

Besides the making of a tracker who can pick up on the song of the track, the questioning way is more challenging and stimulating for me as a guide. It takes much more savvy to come up with encouraging clues and bring someone around to questioning mode than it does to just give a good answer.

"What was unique about the den?" I ask.

He had already described its location, which I knew to be an animal crossroad, so from that—not to mention the sparkle in his eye—I knew there was something he was really itching to tell me.

"What looked like a Coyote track—notice how I said 'what looked like'—and what looked like a Bobcat track, crossed on the sand in front of the den. The quote-unquote Coyote was walking down the road and the quote-unquote Bobcat crossed the road. I couldn't figure out why they were going in two different directions."

"How do they differ as hunters?" I ask. "What about going back to the den and gaining some perspective on the area to see how their trails might fit with their diets?"

I knew there were some open potholes lush with greenery on either side of the road and suspected that the Bobcat was crossing the road to get from one to the other in hopes of coming across a browsing Snowshoe Hare. The Coyote, on the other hand, could've been following the Deer trail that the logging road had become. Rick is sharp.

And motivated. The odds are good that he'll either pick up on these things without any further clues from me or he'll go back again, and again if need be, until the song of the track becomes clearer to him.

Rick doesn't respond right away. "There's something about that section of woods that just doesn't feel right," he finally says. "It's almost depressing. I feel a cloud come over me when I walk into it. I can't put my finger on it, but it's like there's something missing. I know the old growth was cut off and the Pines were planted, but that's not it. There is something else unnatural about it—something that makes me feel nervous, almost suspicious. It's the way I feel when I hop in my car and I know something's been messed with, but I just can't tell what."

"That shows some sensitivity on your part," I reply. "On the surface, yeah, it seems to be a pretty normal stand of Trees. In fact, a lot of folks would feel comfortable there because it's so park-like."

"That's it!" exclaims Rick. "There are Trees, period. Nothing else. And only one kind: Red Pine, all one age, all one height. All you see on the ground are dead needles. It's bare like a desert—no grass, no ferns, no bushes, nothing!"

Looking out over this landscape reminds me of how important it is for the evolving tracker to keep in mind that single lessons can have broad application. I learned the lesson one overcast day in a very different forest from this one—it was all Maple. But how different was it? Both had rust-colored blankets, only one was made up of leaves and the other of needles. Both were a challenge to track through, especially with freshly fallen leaves or needles.

I discovered that it gets easier under certain conditions, such as this day in the Maple woods when several inches of heavy snow matted down the fluffy leaves and then melted away, leaving a slick surface. Coming over the crest of a small hill, I caught the sheen of low sunlight reflecting off of the downslope. Cutting diagonally across the hillside was a jagged line of slightly disheveled leaves. The pattern told me it was made by a Deer, and the assuredness and sense of direction shown in the track said it was the young buck whose territory included this part of the woods.

When I stood up tall, the trail disappeared along with the sheen, because I altered the angle at which I caught the reflecting light. However, when I stooped to again catch the light at the right angle, the trail reemerged.

I'm using the technique right now in this Pine forest and it is working very well.

Rick could use some direction, so I ask him if he sees the barrenness of the forest floor reflected in the animals as well.

"Oh yeah," he replies. "It looks like they just pass through here."

"I'd like to suggest that these observations might be just a reflection of what bothers you. Here's a clue: something is missing, stripped away by a traumatic event."

"Was it done by humans, or was it natural?"

"I forget."

"I didn't expect an answer, Tamarack. I guess I was just thinking out loud."

"I'll give you another clue: look along the edge of the grove where it drops off into the potholes or the bog. The song of the track is loud there."

The next night at dinner, Rick tells me he was out at the Pines that afternoon. "I scouted the edge of the grove, like you suggested," he says, "and it was just so lush! The potholes, too, were choked with all kinds of berries, herbs, and grasses. And stumps. Then it occurred to me that there aren't any stumps in the grove! What was it? Did a hot forest fire come through and burn everything off the highland?"

"The lack of stumps is another voice in the song," I reply. "And a hot fire could have erased the stumps. But there's something else missing. What do you often notice along with stumps?"

"Maybe the glacier leveled that spot," he suggests.

"Sounds like you're fishing for an answer. Remember that to know something is to come into relationship with it. Your answer is your relationship with that grove. When we're in relationship, we don't have to ask for the fruits of relationship, because we already have them. If you'd like to go visit the area again, I can give you some suggestions to help with your relationship."

"I'm on it," replies Rick. "What can I do?"

"Walk through the edge of the grove and allow yourself to merge with it. Move in rhythm with its movement. Remember, everything is alive, everything is walking its life journey. When you move without a purpose, without a goal, you can come into harmony with the movement that surrounds you. This movement, this track from long ago that trails right through the now, is your connection with the past. Like any string of tracks, it can be read anywhere along the trail. When you can walk in those tracks, you will know where the tracks have walked. You will intuitively know everything that grove of Trees knows."

"Okay," Rick says, "I'm back out there tomorrow."

In fact, we both end up out there. It is late in the day and we meet up at a rendezvous site in a Maple grove to head home for dinner.

"I spent the afternoon around that Pine grove, just feeling," he says. "It helped to move because it broke my stare. I had to watch where I was going, and the view kept changing. What I really liked was that it broke my mental focus. I couldn't stay on any one thing. It made me feel alert—I didn't drift off like I usually do when I'm sitting. After a while I'm pretty sure I was observing and feeling more than thinking.

"About halfway through the afternoon, I started to feel out of sorts—something was grating at me. I looked into the grove and it looked artificial, like a cornfield! I bet the area was leveled, maybe bulldozed or something. The only difference I saw between this and a cornfield is that this one grows paper instead of tortilla chips."

"You got it," I reply. "That's the first chapter of our mystery. Now to solve what's missing, besides the understory, that is."

"The topsoil," says Rick. "They scraped off all the topsoil—all the richness—and pushed it over the edge. That's why there's hardly anything growing under the Trees, and why the banks are so lush. Ah—I bet that's why the Bobcat was so attracted to the potholes. The greenery lures the Cat food—Bunnies—which lures the Cat."

"I believe there's one more thing that made you feel uneasy when you first walked into the grove. Something got scraped off along with the topsoil and adds to the park-like feel."

"If it's not stumps, what is it?" he asks.

"You're so warm that it must be giving you hot feet."

Rick looks down at the tip-up he is standing on. It is a mound of soil pulled up by the roots of a toppled Tree. The roots eventually rot away, leaving a mound, along with the depression the soil was yanked from. Over time the mound erodes down and the hole fills with debris, only to be replaced by another tip-up. A natural forest floor is very un-park like—you're continually going up and down, up and down over old tip-ups.

I ask Rick if he has ever noticed how these endless waves of Earth give a forest a unique quality that a prairie or a marsh or a desert doesn't have. You're forced to walk differently, I go on, and it can inspire you to look at time in the way a forest does. Her life is measured in centuries and eons, where the span of our life is mere years. In those special times when I merge with the forest and feel her long, slow heartbeat, I see her tip-ups moving like the waves of an ocean. One tip-up rises and falls, and another rises and falls in its place, undulating the forest floor just like swells on the water.

"Do you see what else got scraped away along with the topsoil?" I ask Rick.

"The song of the forest," he replies. "It was hard to pick up much in that Grove."

"What was hard to pick up? What could the waves have told us?"

He reflects for a bit and replies, "History."

"That's it," I state with a satisfied smile. "These waves are the tracks of times gone by, telling the story of the groves that stood and the storms that felled them. With each crest I mount and each trough I dip into, I read another page of the story. It's like fast-forwarding a video taken over time and watching Trees rise and fall and others rise in their place. Only Trees can be old or young; a forest is both. Only Trees stand still; a forest is moving, always moving."

10
Becoming Wolf

Ready to Learn How One Animal Appears in the Tracks of Another

"Why are you stopping?" asks Rick as I put the van in reverse.

"It's the track we just crossed," I reply.

"Yeah, but you usually don't stop."

"There's something unique about this one," I reply. "It's calling me back."

When my senses are keened and my mind is uncluttered, I can often pick up on the chat of road tracks while driving. When the voice isn't clear to me, I'll slow down and listen more closely, and a whisper here and a glance there might give me the story. I used to have to regularly stop, take time to attune to the site, and then go ask Spider and Blueberry in order to find out the who, when, and why. Now I'm more often able to show up where I'm supposed to be at a respectable time, though there are still those stickler trails that insist I have more to learn. If only those expecting me at three o'clock understood I wasn't given any choice.

Rick, who has assisted me for ten years with the aboriginal skills program, is heading out with me to our camp in the wilderness. The old logging roads that take us to the trailhead are a great place to find tracks, as they're sandy and seldom used. Our route is especially rich in tracks because it crosses a lot of edge—fertile areas where two different habitats overlap, such as wetland-highland and prairie-forest. They're high-use areas for animals because of the shelter, food, and trailways they provide.

"It looks to me like a pretty regular Deer track," says Rick.

"It's not the track per se," I reply. "It's the voice of the track. This track is alive, talking to us. A capable tracker can sometimes hear it even when blindfolded."

"Can you?"

"That's good—always be as a question," I reply in mock seriousness. "And gee, I wish I could answer it, but that would be unethical. As you know, discovery comes from personal experience."

Rick's expression shows that my levity catches him off guard, but he quickly recovers and gives me a smirk.

By now we're out of the vehicle and surveying the tracks. "Let's start here with the footprint part of the trail," I continue in real seriousness. "Remember that it emits an energy, a voice. And keep in mind that it's only one of the voices that make up the song of the track. If we were to get analytical with the footprint, we might miss its voice. Instead, imagine yourself hovering above the Deer as she moves."

Rick looks around to familiarize himself with the path the Deer took, and then he closes his eyes to create the scene.

"Okay," I say when he opens his eyes. "You now have perspective. Go ahead and envision the Deer scooting across the road and making these tracks. Feel the tension that is dictating her decisions and movements. Glance at the trail of footprints and get a feel for what is between them. There is the pattern of her stride, the length of her leaps, and her changes of direction. Here's where you'll see her inner feelings. Can you feel it? Every twist and push is a life-and-death decision."

If there is any great "secret" to impart that will make Rick a tracker in the true sense, he has just heard it. The invisible track—what lies between the footprints—is the lingering presence of the animal's soul. It can be picked up in an instant, and here is one of the reasons the traditional tracker can move faster than the animal laying the track.

"Imagine you're given a picture of a group of people playing a ball game," I continue, "and you want to know what they're playing, and how the game is going for them. You could focus on an individual to see what game she is dressed for, what kind of playing gear she has, her body language, her facial expression, whether or not she's sweating, and so on. Or you could hold the picture back to take in the whole scene and get a feel for the relationship between the players and see what kind of playing field they're on. This second approach is similar to that of a traditional tracker.

"In order for this to work, you not only need perspective, but you have to get out of yourself and into her. She has a sense of self and way of being that is very different from yours. You can't stand up on two legs and look over at someone on four legs and expect to know who she is and why she does what she does. You just came from a house and hopped out of a car, and she has the sky for a roof and goes everywhere on her own power. It is her world you must enter and her pulse you must feel. There's a mystery here. This Deer got scared and bolted across the road. But why? The bog is wide here and there is good visibility in both directions. She has crossed the road many times, so what reason did she have to all of a sudden panic?"

"Could it be she was scared by a car coming?" offers Rick. "Maybe *we* spooked her."

"Maybe," I reply. "However, it's quiet, so sound carries easily. She had time to adjust to our coming, but instead we see this on-the-spot panic reaction. How fresh do you think these tracks are?"

"Very."

"I hear you talking and not the Deer."

"Then I'm confused," says Rick. Showing uncertainty is quite a switch from his old award-winning schoolteacher persona, where he was expected to project an aura of confidence and knowing.

We get back in the van and drive on. After a quiet moment, Rick asks, "What did *you* see?"

This reminds me of the time I was asked the same question twice within the same week. Once it was Rick, actually, who came along with me on what he calls a tracking run, and I stopped to ask him, "Did you just see where that Bear hopped on the Deer trail to cross the bog?"

"No," he replied. "You're going too fast for me—I can't catch much."

"Well, I'm missing a lot too," I said, "but I'm envisioning like crazy and the song of the track is serenading me. You don't have to read the story when you can feel it from the inside out."

"What do you mean?"

"I guess it's like a slideshow, with pictures one after another exploding into my mind. Have you ever heard a song and it triggers images that just start pouring in, seemingly out of nowhere?"

"I think so."

"Well that's what it's like. It's beyond my control. I just let it happen."

The second time was a few days later on a tracking run with a visitor to the yearlong program named Tammy. She told me she had been studying with a Zen Buddhist master and had gained a profound sense of awareness and mastery. "Hey old man," she said, "I'd like to go out tracking with you—I think I can show you some things."

"I think you can," I replied, and that afternoon we headed for a Hemlock forest where I wanted to check out the Deer and Bobcat activity.

After a short time of crawling through Balsam Fir thickets, slinking under deadfalls, and running up and down ridges, all she could say while trying to catch her breath was, "Okay old man, you win."

How could I tell her that it is not about who is faster and better; it is about relationship? How could I help her see that the competitiveness pounded into her as a child had defeated her before she started, and would have still defeated her even if she "won"? How could I get her to see that shortly after my feet touched the needle-carpeted forest floor, I became invisible to myself, forgot about her, and entered another world?

I felt a deep sadness and my heart opened to her, but as near to me as she was, I couldn't reach her. She left angry and confused, which left me hopeful. From confusion comes clarity, and I knew that with her passion, there was a good chance she'd turn her anger into the driving force of her awakening. The next time we go out tracking together, I have a hunch we'll be listening to the same song.

I'm thinking the same could also be true for Rick. I answer his question about what I noticed by telling him that it was more or less the same thing he caught. "It's more about what I *did* with what I observed," I continue. "Or more accurately, it's what I didn't do with it—I didn't try and figure it out."

"Then what did you do?"

"Nothing yet."

"Well, what would you do if you were going to do something?"

"I'd become the Deer. I'd go back in the bog and approach the road just as Deer did, with the same perceptiveness and memory, the same motivation, the same posture."

"Then why didn't you do that?"

"I am."

"Huh?"

"I'm envisioning it as we speak. Just as you don't have to walk in your lover's footsteps to know why she moves and smiles as she does, neither do I with Deer. I'll show you—I'll bring you along with me on this envisioning. But first I must tell you that the road means little to me—that is, me the Deer. You see me from the perspective of the road, your trailway. I travel a different trailway, which you must walk to know me. You'll struggle forever trying to make sense of my actions if you keep viewing them from your world. All you'll know is an illusion, a human construct. Studying those tracks of mine on the road won't get you much closer to knowing me than watching an Earthworm squirm out in the hot sun would get you to know her life in her cool, moist underground world.

"To walk my trailway, just relax, forget who you are, and let yourself be me. I'm coming up to the road and I have to duck low to get through the overgrowth, which irritates me every time I use this trail. Just before stepping out in the open, I need to have my head up, ears perked, and a clear line of sight, to check for potential danger. However, I'm accustomed to feeling anxious here, as I've been using this trail ever since I was born, which is maybe three turns of the seasons now. And I've adjusted to the situation. I often stop just before the brush to check things out as best I can, and then I'll double-check when I'm coming out onto the road. I've always made this crossing without event, and still I never let my guard down. The sad experiences of several of my clan mates have taught me that one lapse is one too many.

"Just as I poke through to the road and raise my head, I pick up a scent that's new to me. Yet I know it—deep in my ancestral memory I know it. I go blank and something inside of me screams, "Run! Run now!" Turbocharged with adrenaline and senses razor-sharp, I take a quick, high bound out onto the road. For an instant, I freeze—I can't see the source of the scent, there's no breeze to tell me what direction it's coming from. Where do I run? I take a diagonal leap across the road to where the trail continues. Only in mid-flight I realize that I overshot the trail and I'm going to crash into the brush beside it. Rather than panicking in mid-air and trying to correct my course, my hard-earned lessons from my frisky fawn days save me. I come down on my back feet first, dig my dewclaws into the gravel, and shoot off to the side and onto the trail."

The next morning, the first words out of Rick's mouth are, "Tamarack, guess what happened last night?" He didn't have to say anything—the sparkle in his eye and his animated gestures told me that something exciting was up.

"Ah, what did you see?" I ask.

"It's not what I saw; it's what I heard."

"As I said, what did you see? To a native, the picture is formed by all the senses working together. Remember, the song of the track is made up of 'voices' of all kinds, including vocal, visual, intuitive, impressions, and memories."

"Yeah, I know. It's just the way I was raised—you see with your eyes, you hear with your ears. All right, I 'saw' Wolves howling last night, just a little before first light."

"Wow—from what direction?"

"Across the lake to the south, and not very far away."

"On the snowmobile trail perhaps?"

"Could be. They were close."

"How many were there?"

"It sounded like at least three, maybe more. Two howled deeply and the others were a little higher pitched."

A knowing smile came over my face, and then over Rick's. Two winters ago the pack ventured out as far as the snowmobile trail and started marking it as the perimeter of their territory. When the snows melted, they disappeared, and now they're back!

The snowmobile trail crosses the road just a stone's throw from where the Deer trail that Rick and I stopped at yesterday meets the road. It was obvious to me when we checked out the Deer's tracks that she spooked because she picked up fresh predator scent. One of the Wolves, probably the male, had ambled down the road from the snowmobile trail to check out the Deer crossing, and he marked it before he left—a standard Wolf practice.

When I lived with Wolves, I'd watch how deliberately the alpha male would choose a spot to mark. He'd spray as high as he could, to impress other animals with his size, and because the higher the mark, the farther the scent carries.

As my neighbor Ben is fond of saying, *Wolves are furry Pigs.* He likes to point out that Wolves eat the dung of other animals, they roll in anything smelly—including dung, and they especially like to smear their cheeks with rotten Fish—the juicier the better. What they're after is to give others a larger-than-life olfactory impression of themselves. This might not fit with everyone's image of the Noble Wolf. Yet if we could step out of our world and into his, we might be able to see the intrinsic beauty in his ways. And the intrinsic necessity.

This story is a good example of how vital a role Wolf's sense of smell plays in his life. And it is the same with Deer. Both have long muzzles, to give a lot of room for olfactory receptors. In comparison, Cats have short muzzles and a poor sense of smell. Unable to scent track like Wolves, they rely more on sight to hunt. They are ambush hunters; the thrill of the trailing and the chase goes to Wolf.

More than anything, this adventure with Rick, Tammy, Wolf, and Deer was a lesson for me in how our illusions can cripple us. We make what is easy complicated by trying to fit it into our preconceptions. Through envisioning, I've learned that I can make things easy by letting go of notions and taking things for what they are. To track effectively and consistently, we need to see what *is,* rather than what we think we ought to see.

Tracking is truth. The more honest and forthright we live, the better we are at tracking. We are creatures of habit. If we are often insincere and have trouble grasping reality, we can't help but carry these patterns into the woods with us. They will trip up our tracking the same way they trip up our lives.

A tracker who can trust his intuition and senses could make a trusted friend. If he can hear the song of the track, he can hear the voice of my heart. If he can become an animal and attune to his feelings and desires, he can be sensitive to mine. He could be an example for others and an asset to his community. If in my final hour I can look back at my life and see that I have become a fair-enough tracker, it might also mean that I have become a fair-enough person, and I will have considered my life worth living.

11
Old Songs Never Die

Teachings from the Ghosts Who Linger in an Old Forest

It has been a century or more since the ancient Trees of this forest were felled, and it looks as though she has recovered beautifully. Her stately, vigorous White Pines and Sugar Maples are in their young adulthood, and many others not so long lived, such as Birches, Aspens, and Firs, are well into their elder years. In fact, some already lie on the forest floor, quickly becoming the rich humus that will nourish the next generation. To most people hiking through the area, this probably appears to be a healthy, natural forest.

It is and it isn't. Everything leaves a track, and the song of the loggers' track bellows out at me this afternoon as I check the area to see if the Wolves have come around lately. A virgin forest has a certain feel—it encourages trust and openness, and I find myself naturally relaxed and curious. In a forest that has been logged, I sometimes grow suspicious and wary. Even though there may not have been a human through the area in decades, the spirit of the loggers lingers. The forest just doesn't feel right—her balance has been disturbed. It takes a lot longer than a century—a lot longer than a single generation of Trees—for the balance to be restored.

I believe the impression of a healthy forest—a healthy environment of any kind, for that matter—is imprinted in my genetic memory. It is a survival trait that told my Ancestors where they could dwell and forage in peace and prosperity.

When the loggers came, many of them said, "Look at these Trees—they're past their prime; they'll soon die and go to waste. They should be cut and used." Now the forest managers come and look at the young, second-growth forest and say, "Look at how tall and vigorous these Trees are!"

Yes, they look good. So where is this intuitive feeling of uneasiness coming from?

It is the sterility.

The lush, rapid growth is an illusion—a scab over a scar, a forest trying to heal herself. The wound is much deeper, much more critical to the life of the forest, than can be healed by merely replacing the ancient Trees. Along with them went the Moose and the Marten, the Fisher and the Wolverine, the Ginseng and the Orchid, the Grayling and the Sturgeon. Gone are the old snags, the great standing hulks of the old forest giants where so many Birds and other animals once nested and denned and foraged.

The forest floor is now clean and park-like. Gone are the shed branches, some of which were as big as the Trees that now stand. And gone are the fallen Elders who used to provide cover for ground-dwelling animals. The rotting trunks provided the nourishing seedbed some Trees needed for their saplings to get a start. A forest like this one of even-aged, fast-growing Trees occurs naturally only when there is a devastating event such as a major forest fire or wind storm. And then the young forest still looks and feels different from the regrowth after logging. Fire leaves charred hulks standing, and the forest grows rich from the Trees' mineral ash. Animals and plants return quickly from beyond the burn.

Like scar tissue, these young fast-growing Trees aren't as strong as those they replace. Their wood is soft because it grows fast. The annual growth rings of old, slow-growing Trees are close together, which gives these Elders strength. The same is true of the roots: quick-growing Trees topple easily in the wind, whereas the Elders have gained resilience from standing in the face of many storms.

A healthy forest is a sick forest—it is filled with diseased and dying Trees. As contradictory as this might sound, we need to grasp it so we can learn to back off and let the forest be as she is supposed to be. Without sickness and death in the forest, there cannot be life and health. Ailing Trees with hollow, rotting trunks provide nourishment and living space for a host of life forms. Woodpeckers, Squirrels, Porcupines, and even Bears rely on them for shelter and places to raise young. Parasitic grubs and caterpillars feed countless Birds, amphibians, and reptiles. A second-growth forest has little of that to offer.

And here I stand, feeling this—feeling uneasy. This uniform lushness is an illusion, and I can sense it in the same way that I know the omnipresent death and rot of a natural forest as bursting with life.

The song of the track never dies. After one hundred years, the logger's ax still rings through the forest. The sweating Horses can still be heard snorting as they drag their bobsleds laden with immense logs to the head of the rail spur. The *choochoochoo* of the locomotive's escaping steam still lends rhythm to the *shhhhh* of the Wind in the needles of the high Pines. The old song joins in the chorus of the one following it, and this continues as one great, unending symphony evolves.

The more the songs merge, the harder they are to distinguish. But not impossible. What is a Wolf but the sum of all the Deer and Mice he has eaten in his life? What is the lodge but the sum of the bark and root and limb that have been put into its making? Each piece of bark can be tracked back to the Tree who gifted it. And each fiber of a Wolf's muscle can be tracked back to the Deer who offered it. The only limitation in tracking the bark or muscle fiber back to its source is the tracker.

Right now, for example, I am crossing from one side of a valley to another by taking a razor-back ridge. Or so it seems. This ridge didn't exist a hundred years ago. Before I was attuned to the song of the track, I probably would have thought the ridge was here just as long as the hills it runs between, which were sculpted ten thousand years ago by the glacier, and have hardly changed since.

As I walk along the ridge, I feel a tug that draws me to look down and to my left, where I see an odd-looking pond that the glacier could hardly have formed. There is where the soil was dug up to build this ridge. The ridge itself supports Trees just about as old as the forest around it. However, the ridge Trees don't feel old. The rest of the forest floor is pockmarked with stumps and mounds left from the root clumps of rotted-down Trees torn up by long-ago storms. The surface of the ridge is almost perfectly smooth, and its steep sides are nearly perfectly parallel to each other. Were it not for the Trees, I could walk the top of this ridge as easily as sauntering down a sidewalk.

One hundred years ago, the Pine that was logged from the North

Country was floated down rivers in great springtime logging runs to sawmills, where it was cut into lumber to build the bustling cities of the Midwest, such as Chicago and Milwaukee. Many of what look to be old Beaver dams on the area's streams were constructed by loggers to back up water for floating logs down to the large rivers. Even though those dirt dams, often hastily and crudely built, look more natural than this ridge, I still get an uneasy, anxious feeling around them, as though something is not quite right. When I look closely at the dam, I don't find any wood, which along with dirt is the Beavers' primary building material. And I don't find a trench. Beaver get their dirt directly from the upstream side of the dam, which leaves a trench adjacent to the dam. The log drivers, on the other hand, usually got their dirt from a highland site.

The days of the river runs were from an era before the loggers who built this ridge I'm standing on. These loggers were after the big Sugar Maples that grew on the fertile highlands. Maple is a dense hardwood that doesn't like to float, so logs had to be taken out by rail. Short spur lines were laid throughout these hills to reach the Maple Groves. Small locomotives that were built specifically to haul logs ran on narrow-gauge tracks, which were hastily laid and pulled up as soon as groves were logged out. The railroad's brief presence is permanently etched in the land by the scars of altered topography such as the one I'm walking on.

Coincidentally, my family has a century-old log Cabin that was built by the Thunder Lake Railroad Company—the logger barons who stripped these hills. The Cabin was the first one to go up on the shores of these wilderness lakes. The company built it to entertain their clients, who they would bring up in a special passenger car. From the Cabin they would go canoeing, hunting, and fishing. Soon the trains were bringing wealthy passengers, such as Buffalo Bill Cody and the Pabst Brewing Company family, up to their own cabin retreats.

I wonder what it was like to stay in a log cabin in the middle of a forest of knee-high stumps. Clear-cuts sadden me, yet if I were living back then, I might have felt like many others and seen them as signs of progress. The wilderness was endless, and taming some of it for farms and towns was thought to be a good thing.

A few of the old stumps are still here to tell their story, and the few pictures that I've seen from the era show a forest leveled from horizon to horizon and right down to water's edge. Were it not for the stumps, you'd think the pictures were taken somewhere out in the Great Plains.

Some of the song of the logger's track is the song of our Cabin. As Wolf is Deer, so this Cabin is the forest. I can listen to the song of one of the Cabin logs and learn about the grove from which she came. She will tell me her age, how many neighbors she had, how close they lived, whether she grew on high or low land, and what the weather was during her youth.

The original roof was probably of rough-cut lumber, sawn by a steam-powered mill. Cedar shakes were hand split on site. Judging by the age of the planed tongue-and-groove boards on the present roof, the original lasted about fifty years. I would guess that because of the remoteness of the Cabin and its sporadic use, along with the changing economy, the original roof wasn't well maintained, so it started to leak and rotted away. As goes the roof, so goes the floor, which the replacement floor confirms.

Fortunately, the fate of the roof doesn't necessarily seal the fate of the building, especially if it is a log cabin. I know of a couple of old trappers' cabins back in the wilderness whose walls have stood another seventy-five or eighty years after their roofs and floors have returned to soil. The logs were resilient because they came from old-growth Trees that grew slowly. That gave them a good core of solid wood, which is called heartwood. It has mineralized and it is often saturated, which keep it from rotting. Fast-growing Trees are mostly new growth, or sapwood, which rots fast.

Something doesn't have to move in order to leave a track. A log cabin is a living entity made up of living entities, as much as is a Wolf.

Sometimes I'll come upon an old cabin in the distance and I'll hear its song. It'll paint me a picture of the cabin and her surroundings as they were twenty years ago, fifty years ago, one hundred years ago. This'll help connect me with the songs of the people who built and lived in the cabin. I'll watch them peeling the logs in the clearing and setting up tripods and pulleys to hoist the logs one atop the other. As this song fades, I might pick up the song of the primeval forest before any Euro-Americans set foot in the clearing.

Back in the days of the first trappers and missionaries, it was the custom to leave cabins unlocked. They usually weren't lived in full time, as people were out and about on their business. They traveled long distances, and settlements were few and far between, so if someone came upon a cabin and needed shelter, he was welcome to use it. If he needed food, he could help himself, and if he had food to spare, he'd leave some. He'd leave the cabin in as good or better shape than he found it, and he'd usually chop and split some firewood for the next folks to come along.

Our cabin was different—it was owned by lumber barons who lived in the city. They weren't part of the wilderness tradition. Their idea of security was a door with a padlock and iron-barred windows. They apparently didn't know that it was acceptable to break into a cabin if necessary. In the city, a locked door might deter someone, but in the wilderness, there is no one around to hear a door being kicked in. Judging by the type of wood and style of hinge used on the replacement door, the original door was probably destroyed in the first decade or two of the cabin's life.

Shortly after that, a cooking addition was added to the cabin, which tells me those were probably prosperous years for the Thunder Lake Railroad Company. It was the Roaring Twenties, and they were truly roaring through the forest. By the end of the decade, the timber was gone. The country sunk into an economic depression and the denuded land was abandoned and it reverted back to federal ownership.

Fortunately for us, an enlightened president gathered up the tracts and created a national forest. The railroad pulled up the last of its tracks, and what was to be our Cabin stood abandoned. It is probably during that time that the roof and floor rotted.

World War II restored the economy and the working class started to get a piece of the pie, which kicked off the Northwoods' modern era. The Northwoods were a strong draw for urban folks whose childhoods were filled with stories of Indians and fur trappers, voyageurs and missionaries. They got around in Birch bark canoes on picturesque lakes with big Fish in the pristine waters and Eagles soaring overhead. Bear and Moose roamed the shadowed forest.

Lakeshore lands were subdivided and little three-season cottages sprang up nestled in amongst the vigorous new-growth Trees. It was during that time, the late 40s and early 50s, that the Cabin was purchased and a new roof and floor were installed.

As I walk this ridge immersed in both the song of the forest and the song of the Cabin, I realize there is only one song. There is a resonance

between Cabin and forest, for they are literally each other. The fates and fortunes of the two have been intertwined from the beginning, and they continue to be. This is the way of the forest and of all who dwell there. The forest gave of her Tree people so that the Cabin could come into existence in the same way the Deer people give of their kind so Wolf may live. This Cabin, which embodies the spirit of the forest, now serves as the office and gear shop for the Teaching Drum Outdoor School, where students come to learn how to live in balance with the forest and to be her guardians.

Continuing across the ridge, I hear the song of the loggers' track returning. It permeates my being and I feel uneasy, as though I'm being watched. Over there to my right is the old Moss-covered stump of a once-towering five-hundred-year-old Pine. The two lumberjacks who are felling the Tree with their double-bitted axes could look up any time and see me walking the railroad grade. I turn around and there is a man with a pickax cutting into the side of the hill to level out the Horse path so that the log sleighs won't tip over with their top-heavy loads this coming white season. The wielder of the pickax is little more than a boy. Judging from his dress and the way he moves, he looks to be a farmer. He probably has enough brothers and sisters to help on the farm, so his father sent him along with the crew recruiter, to help build his character and earn some needed money for the family. He works steadily, with his full focus on his task, so he only gives me a glance.

These signs, these voices that contribute to the song of the loggers' track, are softened with age. Their rough edges are smoothed with the effects of weather and animals. The scars are mellowed by overgrowth and the accumulation of forest duff.

If I were locked in the present and didn't hear the song of the past, I might not recognize that any logging occurred here. The present would be whatever my senses gave me and however my mind interpreted it. However, the present is not only what I perceive in the now, and my understanding is not the understanding of the forest. The song of the track breaks down these barriers and takes me into the consciousness of the forest, the timelessness of the now.

In the continuum of the now, past, present, and future coexist. The consciousness of the forest still includes the glacier and the loggers, along with what is to come. The future already exists for the forest, because she has no concept of time. She doesn't ponder, she doesn't wonder, she doesn't project. She just is. This is the consciousness in which we dwell when we're immersed in the song of her track. And this is why I experience the forest as she was when the loggers were here simultaneously with the way she was before they came.

Winding my way through the Aspens and Firs, I realize that I'm feeling anxious, even suspicious. It is not just because of the time warp I'm in or the human presence from the past, but because this human activity is out of balance. We have always been a part of the wilderness, right along with Cougar and Porcupine. However, the human presence that I now sense doesn't allow me to feel that they're here to be a part of it—they're here only to take.

Rather than dwell on the feeling, I become the animal I am, snaking my way through the dense Firs and Aspens who have replaced the steel rails and Tamarack ties. It frees me of the shackles of past and future.

I pass by a two-year-old Porcupine, who is head-high on a Tree trunk about five paces away. She climbs a couple of her body lengths higher and looks down at me. If I could be fully present with her, we could develop a relationship. Perhaps we'd talk, telling each other of our journeys. If I took the next step in awareness, I would become Porcupine. I'd then see how she sees and feel how she feels. Having her memory, I'd know where she came from, and having her aspirations, I'd know where she was going.

This illustrates one difference between reading a track and hearing the song of the track. When I read the footprint, it speaks to me, and this is beautiful. I'm getting to know the track maker by conversing with him through his footprint, in the language of the footprint. I could also become the animal, which makes me the print maker. The print itself now takes on secondary importance, because I sense and feel directly the reason the print was laid. It would be the same as if I were walking out to the lakeshore to push off in my canoe. I'd already know the reason for my movements, so studying my own footprints to find out would be redundant. Instead of going over the past, I could already be out fishing.

Here is one reason some native trackers are so effective and why their tracking can seem effortless. Some of us modern folks have made a science of tracking, and science means study, analysis, data, and processing. For the native, tracking is being. There is instant access to information. There is intuitive clarity. There is little question because there is the knowing that comes from being.

For about a half mile, I continue following the grade. Sometimes the glare of the worn steel rails in the dappled sunlight catches my eye; however, I pay it no attention. The rails are long gone, yet they remain as a voice in the song of the track. I was taught that it is impolite to stare, and in this case it would also be harmful, as the rails would disappear. Along with them, the song might disappear as well—or at least change. When the song is made up of a chorus of voices, we can change the song by messing with just one of those voices.

Through the undergrowth I see an area of brightness just ahead— a small clearing. I pause. In mid-afternoon at this time of year there should be some sign of activity, but it is unnaturally still. I instinctively abandon my identity and become a Tree trunk.

And none too soon. In the same instant, a human form emerges from the woods into the clearing. He is too far away for me to make out features, but I know the posture: it is Jacobs, a staff intern. He, along with Kip, a student who is volunteering between courses, came out with me to go fishing while I did some scouting. They dropped me off about a mile before the lake. I thought they'd have been fishing already rather than traipsing through the woods. Maybe they came over this way to dig bait Worms from the sweet, rich soil under the nearby Maples. Knowing them, I bet they took a lunch break of Spring Beauties, Violets, and Sweet Sicily while they were there.

Honoring the state I am in, I stay put as a Tree trunk. Was Kip ahead of Jacobs or behind? Or maybe they split up and Jacobs is alone. He is an independent sort, so I wouldn't be surprised. Kip is impetuous, very much a creature of the moment, so he may be off on a little side adventure. A Tree trunk knows no time, so has no trouble waiting to see.

I become the Thrush singing from the top of a distant Maple; I become the Mosquito who is biting my temple. At the same time, I remain sensitized to any voice that might attune me better to the song of the track. I'm at one with the forest: ever present, ever observant. What I am not is reactive, nor do I filter or judge what my senses and intuition bring me. They're not mine to tamper with—they are the voices and feelings of the forest I have become.

I'm able to do this only because the Tree trunk and I are willing to let go of who we are and trust in who we are a part of. We are one and the same, each of us an organ within the greater organism of the forest, each contributing our gifts for the good of the whole. It is this shared resonance that allows us to seamlessly shape-shift into each other.

Is it easy? After you've done it once, it becomes part of your makeup. That is because it was always there; it just had to be rediscovered. It is our sense of self that has kept us from it, and when we can leave our self-consciousness behind, our inner balance becomes apparent. We then naturally dwell within the balance of the forest. There are no boundaries, no distinction between self and other. It is the ego that defines us as separate from the rest of life. To become the forest, then, all we have to do is be what we already intrinsically are.

Everything is pulse and rhythm—a never-ending expansion, contraction, expansion. It is this rhythm, or I should say attunement with this rhythm, which allows the crossover, the shape-shift. Within this rhythm, increase is decrease, the hunter is the hunted; to be full is to be hungry. You can expand beyond yourself or contract within yourself at will, because those of the forest are you, and you are those of the forest.

I feel the call to move, to cross the clearing. No longer a Tree trunk, I still remain what I have become—a movement within the greater movement, a forest within the forest. I listen to the song of the track, and I am the song of the track. I sense my own presence in the same way I sensed the presence of the loggers.

In the middle of the clearing, I feel vulnerable. The shadows ahead beckon me, but I don't respond to this impulse because it would project me into my future and I'd lose my full presence in the now. I accept the twinge of fear that vulnerability brings as good. All feelings are for a reason. The fear begs my adrenal glands to release a bit of adrenalin, which in turn coaxes my liver to raise the glycogen level in my blood, which causes me to feel a slight head and body rush. I'm now hypersensitive and alert. I take a couple steps and I feel something, but I don't react. I don't show any outward sign that I might be on to something. It takes a couple more steps before I feel prickly heat on the back right side of my head. Without breaking stride, I turn around and wave in the direction from which I feel the heat. Kip, crouched down beside the trail in a quick effort to hide from me, sheepishly waves back.

My instinct is to keep moving. If the presence were a threat, it would be important that my movement maintain the illusion of composure and purpose. If I were to stop, I could be perceived as a threat; if I bolted into the cover of the Trees, I could invite a chase. Maintaining my movement also helps to maintain my centeredness. I remain in the state of becoming, whereas if I were to stop or alter my pattern, I'd return to self-consciousness.

I learned this from stealth and camouflage training. I'd practice it over and over in a variety of circumstances so that I'd learn to do it automatically. Of course, Kip was no threat to me. Yet I didn't know it was him until I actually saw him. I was already in a state of vigilance and suspicion because of the ever-present possibility of a stranger showing up. As well, I've been trained to be as a question and draw conclusions only when I have to. In this way I maintain maximum flexibility and adaptability regardless of circumstance.

All I knew in the middle of the clearing was that there was some presence focused on me. Had I thought about it, I would have given the odds to it being Kip. However, this is not something that a person in balance would do, whether human or not. And it is not what I was trained to do. Why? Because if I figured it was Kip, I'd have relaxed and exposed myself. If it turned out not to be Kip, I'd have put myself at risk.

Every day is preparation for the next day; every situation is training for the next situation. There is a training period—it is called life. I don't expect my training to be over—and I hope it won't be over—until I take my last breath. Training is growing, training is being humble and ever alert. My experience in the clearing—in fact, the whole day's experience—is training for my next time in the forest, for my next day of life.

The next morning Kip stops by to talk about the experience. Besides tracking being by far his favorite topic, he wants to know what I think about this new tracking method he has seen where you feed data into a palm computer out in the field. "And then the field data is fed into a main computer," he says. "It crunches the numbers and makes a map showing the density and movement of the animals."

"What kind of field data is collected?" I ask.

"Oh, things like location of scat and track, lays or loafing areas, browse and kill sites, trail locations. Stuff like that, the usual sign."

"Let's see . . . collect data, compile data, analyze data. . . . Sounds strangely familiar to me . . ."

"Okay, it's the scientific method, but maybe it is a way to legitimize tracking with law enforcement and game management folks. And couldn't it be a doorway to greater awareness?" he asks by quoting one of my catchphrases, as he looks at me with a smile and a raised eyebrow.

"At least it's getting people out," he quickly adds when I roll my eyes back at him. "It's got to be good to just get people out looking for sign."

"Every time we use a doorway to get what we want," I reply, "it reinforces the need for that doorway. We're creatures of habit and pattern, so if we think we need the doorway, we need the doorway. When we step out of a doorway, we come from the perspective of an observer, looking from the outside in."

"I understand that," he replies, "but I don't see what it has to do with a computer helping us track."

"Of course I don't have first-hand experience with what you're describing, nevertheless it sounds to be yet another effort to adapt the scientific method to tracking. It's still studying the track to know the animal rather than knowing the animal to know the track."

"But you're talking over most people's heads," replies Kip. "How many are going to be able to experience what a native tracker does? How many even *see* an animal when they go out? Some sign is often all people find, and sometimes not even that."

"I hear you," I reply. "And I'd like to make one correction—I'm not trying to talk over their heads, but below them, to their heart-of-hearts. I think the intuitive tracker at the heart of each of us knows what I'm talking about. How would I enter the fear and vulnerability I felt yesterday in the clearing into my palm computer? We're not talking observational data and analysis here, we're talking intuition, sensing, and feeling. And then there are ancestral memories—those knowings from the dim reaches of our evolutionary past. It is information not even in our conscious grasp, yet we respond to it. We're talking about something broader than your everyday body-mind experiences here. Remember, native tracking isn't about being, it is about becoming something other than what we are. In other words, it is not based just on what is, but also on what isn't. It is beyond the consciously known, beyond what we can point to or wrap words and numbers around.

"Modern methods of tracking can be learned—they can be studied and practiced, and we can get a computer to do some of this for us. Native tracking cannot be learned. You don't learn intuition, it's already there. All that's needed is to clear away the rubble so that it can function unhampered. Sense and instinct are not quantifiable. Feeding them into a computer would be like trying to capture the wind in a bottle. Tracking involves skill, yet it's primarily an art form. Being an artist is not just about producing something, it's a way of life. It involves your whole being. The artist envisions before she creates. In other words, she becomes her creation. The same with tracking—we cease to be trackers, we are no longer ourselves when we become the animal. Like the artist and the art, the tracker and the tracked have become one. Calling ourselves trackers, then, is redundant because when we are the animal, there's no need to track him any more than we need to track ourselves."

"I get glimpses of that," says Kip, "but where I get tripped up is how it applies to my daily life."

"I'd like to suggest that you look at it not as *applying* to your life, but *as* your life."

"Now you're talking in riddles," says Kip. "Give me some fat to chew on."

"Okay, how about this chunk of fire-roasted Bear fat?"

"Oh, you know what I like!" exclaims Kip. "Serve me up."

"You bet! Imagine the hunter-gatherer as the quintessential human—fully alive and completely developed. He's adaptable and able to perform the range of tasks needed to not only survive, but thrive. Now imagine walking into the woods and feeling completely at home. You know the plant and animal Relations around you, you're able to understand their language, and you trust that they'll be there when you need them. If this were you, in this state of being, do you think you'd call yourself just a tracker?"

"I get where you're going," replies Kip.

"All right, let's say we're visiting a native camp and we want to meet their tracker. Can you imagine the blank looks we'd get? For sure there'd be people who could track—probably all of them, in fact—but I'd be surprised if any one of them would step forward and say, 'I'm the tracker.' When the women go out gathering, they use their tracking abilities to find ready nut groves and berry patches, the basket makers track down patches of willow shoots, the scouts go out and track down new campsites."

"But specialization can work," counters Kip. "Computer-assisted tracking does zero in on animal activity. I've seen it in action—it shows population density, movement patterns, forage base, lots of things."

"You're certainly right," I reply. "When specialization works, it can work very well. On the other hand, it might just be an evolutionary dead-end. When another skill is needed, the specialist is a goner. As you know, specialization is a civilized trait, and hunter-gatherers are generalists. They wear many hats, not only because they have to but because they're intimately involved in all the aspects of their lives. Most of them can do whatever is called for at the time. A woman going out to pick berries might come across the track of a wounded Deer and end up bringing in meat instead of fruit. Or her partner might have gashed his leg on a sharp rock and need immediate treatment to stop blood flow and prevent infection. These are all the affairs of daily life, and whether they're expected or unexpected, pleasurable or painful, they are all equally important and demand equal attention. If someone specialized only in berry picking or bow making, they might be able to make great contributions at times, and then at other times they'd be dead weight—even liabilities."

Satisfied with our sharing, Kip and I—generalists that we are—switch from tracker to tanner and head out to the yard to work on some Beaver and Coyote pelts. It's easy here to lose ourselves in the scraping—the rhythmic *shhh-shhh-shhh* is like a drumbeat connecting us with an ancient rhythm. Without trying, we become the pelts, feeling them soften and open up to take in the tanning solution. There is no thought of using modern shortcuts for this craft that feels so deeply satisfying and gives us time for quiet reflection.

12
A Sound Lesson in Tracking

How to Let Ears Become Eyes That Can See the Invisible Trail

As with other mobile beings, Sound's actions are affected by his surroundings. Like Birds, Sound travels better with the wind than against. Sound is not a social creature; he does better alone than with others of his own kind. He is tremendously adaptable—he can travel through air and water, and even metal. However, he is fickle when it comes to weather. He despises snow, but give him heat and calm air and he thrives. If you want to see him at his best, catch him on a warm night over a large Northwoods lake. If you're lucky, you'll catch him carrying a Loon's yodel so well that you'd think it could reach the far corners of the Earth.

When I hear a Sound in the woods, I hear a life. I'm not referring to the life making the Sound, but to the Sound itself. I've come to learn that a Sound has a life of its own, independent of its maker. Once a Sound is created, he is an independent being.

The reason I look at Sound as a living being is that I track. Sound is one of the more important tracking signs, and sometimes it is the only one I have to go by. Yet this doesn't have to be a limiting factor, because a single Sound can say much. The better I understand Sound's language and temperaments, the better the tracker I'll be. If I can recognize Sound as having his own personality, his own passion, and his own sense of purpose, I'm going to take him seriously and give him my attention. When I hear a purr or a scream, I'll attune my senses to pick up every nuance of character and feeling I can.

If I want to be an effective tracker and a responsible member of the Hoop of Life, I need to create Sounds responsibly. It helps when I realize that Sound has a life of his own and I have no control over him once I've let him go. I have to live with whatever he affects and whatever he brings back to me.

We modern people are sight oriented; most of us don't give Sound much attention other than in the form of speech or music. We know Sound mainly as something that travels from point A to point B, and we have little awareness of what happens in between. And wouldn't you know; it is in this in-between area where Sound picks up the traits that speak to us trackers.

All of Sound's characteristics fascinate me, yet what helps me most as a tracker is understanding the way Sound evolves, and Sound's ethereal nature. Although we might see a Sound's source and effect, we won't see Sound himself. Like a spirit, he can pass through solid objects and trigger events so fast that we don't have time to react.

We are nothing by ourselves. We are defined by our relationships, and the same is true of Sound. His relationship to other forms of life gives us an easy doorway to knowing him. Like begets like, so when Sound comes from a living being, he possesses some of the life of that being, and in turn he helps create more life. Who would say that the mating chirp of a Cricket does not produce more Crickets, or that the howl to bring a Wolf pack together for the hunt does not feed the pups who will one day go on to start new packs?

Without living Sound, life would be very different from what we are accustomed to. At this moment, for example, you wouldn't be listening to this story. There would have been no telling and nothing for the tape recorder to catch. Even if we were together, we'd either have sat in silence or evolved another method of communication. Imagine our Ancestors sitting around the hearth with no voice to carry the stories of the hunt, the questions of the children, and the wisdom of the Elders. Imagine having no voice to express the pain of your soul or the joy of your heart. Imagine a world without the clap of thunder and the patter of a raindrop.

At the same time, let's remember that Sound is much more than a one-dimensional auditory experience. Thunder wouldn't have near the power and raindrops wouldn't be as soothing if they couldn't be felt and seen as well as heard. Music has vibratory and visual qualities that add to its enjoyment, and speech is facial expression and gesture along with voice. My deaf friends can hear me right along with those who are not deaf. The difference between those of us with and without compromised senses isn't whether or not we can hear, but rather *how* we hear.

Natural-living people honor Sound just as Cricket and Wolf do, in part because they need to. Sound helps feed them and keep them safe. He brings them comfort and pleasure. For them, Sound is sacred.

Even though you might take much of the Sound in your life for granted, or even consider some of it a nuisance, my hunch is that you still have some sacred relationship with Sound, much like your wild Relations. Even though your relationship with Sound may not be as intimate or as vital to your survival as it is for them, you are still very much like them. Some people's relationship with their favorite music borders on the sacred, and they can feel its life and spirit. Whose heart doesn't warm when they hear the voice of their loved one, and whose blood doesn't curdle when they hear a terrible scream in the night? Doesn't Sound have to have life and passion if he can so deeply touch the lives and passions of others?

This awareness of the power and sacredness of Sound helps the native tracker be a fast and efficient hunter. We give attention to what we cherish, and what we cherish lingers in our hearts and our memories. Here is why the native tracker is attuned to the lingering voices of the song of the track.

Although this may sound like some esoteric mystical experience, it is actually so much a part of us that it can be hard to recognize. Just like the effect of someone's voice can stay with you long after the person has left, the reverberations of an animal's call continue to have their effect well after the call has died out. This is the animal's shadow voice. It will often tell more about the animal than his initial call. A male Cardinal's song can tell me about his health, whether or not he has a mate and if she is on eggs or hatchlings, and so on. Yet his shadow voice—the effect of his call—tells me so much more. Like an echo, it reverberates on through his Hoop of Life and comes back to me through other voices, which teach me things like the size of his territory, his relative dominance in the Cardinal clan, the directions in which he heads out to forage.

If I wanted to trap the Cardinal, the information I garnered from his shadow voice would be a bigger help to me than what I might have learned from his initial call. In fact, I wouldn't even need to hear the call. This is the way with the song of the track: we weren't there when the animal passed through, and still the reverberations linger for us to hear.

I'll tell you about a real-life situation that might give you a feel for how this sacred way with Sound can manifest. It is night-time, and the color of this particular night is black—not dark, but black. Thick overcast and sticky humidity make it feel as though the atmosphere has collapsed and is trying to smother you.

This is the sort of night when a person can see better with his eyes closed. When one of our senses is unable to function, we compensate by sending the energy normally reserved for that sense to our other senses. I welcome this opportunity to practice "seeing" better without my eyes.

I'm sitting around the campfire with several students and decide to leave for home. As I turn around to head for the deer trail leading out of camp, two students named Kip and Amber ask if they can join me for part of the way. As is my custom, I step aside and let them go first. They gain experience in way-finding, and I can lag behind in the quiet of their wake (or go off trail, as I often do).

Out of habit, I had kept one eye closed a short while before leaving the fire, so that it'd be more light sensitive in the dark. This night it makes no difference, I'm finding out, as both eyes are equally useless.

The first half mile of the trail, which rolls up and down small hills and loops around woodland ponds, is fairly easy to follow in the dark, even for someone who has only walked it a few times. Besides the trail being worn in just enough to be able to feel with your feet that you're on it, it rises and drops and veers to the right and left in a rhythmic way that makes following it like going through the steps of a familiar dance.

Not so with the next part of the trail. It comes out of a dense stand of upland Pines and onto a level, open Maple forest. The leaf litter is matted down from the heavy snow cover that just recently melted off, so it is hard for one's feet to distinguish the compressed debris on the trail from the forest floor in general.

It isn't long before Amber and Kip wander off the trail, one to the right and the other to the left. They lose track of each other, which is helped along by the thick and steady drizzle that began right when we came out of the Pines. Following our night-walking protocol, they keep silent, which makes it even less likely that they'll reconnect.

Kip exudes confidence, so it is not surprising that he struck out on his own. Amber, on the other hand, is often gripped by fear. At the same time, she feels the need to prove herself, so I doubt that she'll call out for help.

Many night Sounds are subtle, and our silence clears the way to hearing them. When it is too dark to see each other, silence helps us stay connected, as we can hear each other's movements. When it is quiet enough, it is even possible to feel another's whereabouts. If we talked to stay in touch, we couldn't feel each other as easily and we would drown out the night Sounds.

I'm fine with the two drifting off, as I know they are each in their own way enjoying the challenge. The night is warm and they have both learned enough to take care of themselves until dawn if they weren't able to find their way to the road. My hunch is that Amber sees this as a great opportunity to embrace her fear, and perhaps Kip's self-assuredness will be dampened enough to allow him to gain some perspective.

I can tell direction in the dark, and I can follow the trail using senses other than sight, so I'm not concerned for myself. I drop back and listen for the occasional snapping stick, which gives me their whereabouts until they drift out of earshot.

Eventually Amber makes it to the road and goes back to help Kip.

"How did you stay on the trail?" Kip asks me.

"I just couldn't come up with a good reason to go off," I reply.

"Hmm. I guess it's not impossible to stay on it, but I had a hard time. How did you do it?"

"You're asking the wrong person," I reply. "Ask the night; she was my guide."

I don't have to see their faces to know they're rolling their eyes at that one. They had to see it coming, yet they'll sometimes ask anyway, just to see if they can get something out of me.

"Seriously," I say, "I suggest you go back over the walk you just did. Envision coming out of the Pines and into the Maples. The mother of the night is there waiting for you, waiting to guide you through. Become her, become the night, feel the sensations. What was touching you? What were you hearing and intuiting? These are the fingers of the Mother gently caressing you and guiding you to safety and comfort."

We're all quiet for a little while, and then Amber whispers, "This is cool. It feels different this time—I'm more relaxed. When I came out of the Pines the first time, I was anxious. I felt lost and started groping around. It's odd—I felt fear even though I knew you two were close by and I could call out at any time. But I wasn't going to do that—I was going to do this on my own."

"I got frustrated," adds Kip. "I felt silly stumbling over deadfalls when I knew the trail had to be just a few steps away."

"Okay, you see the difference in your state of being between then and now. Take those initial feelings back into your envisionment. Once again, you're just coming out of the Pines into the Maples. Instead of pushing ahead—instead of forcing yourself into your fear and uncertainty—just stand there, just be. Regain your centeredness, and from there, expand out beyond yourself. Your physical self is no longer your physical boundary. Let the night be your eyes, ears, and hands. Become the night for a bit, and tell me what she's saying to you."

Amber is the first to speak. "I feel the drizzle on my face. There must be a light breeze because it's hitting my left side a little stronger. I hear drizzle hitting the leaves on the ground in front of me, and behind me big drips are coming down off the Pines. And I hear the Frogs in the background. Ah, Tamarack, that's it—I think I got it! I could keep the sound of the drips behind me and the rain hitting my left side, and I'd walk straight out to the road. Ha, who needs the trail?"

"For how long would you be able to hear the dripping Pines? And what if the wind died down?"

"Good questions," says Amber. "I guess I'd be lost again."

"So what's next after the Pines?" asks Kip. "There should be something else."

"Several things," I respond. "The Night extends a strong guiding hand and you're only grabbing on to her little finger. I'll give you a clue—it's something so glaringly obvious at this time of year that you're probably taking it for granted."

"The Frogs?" asks Kip.

He's got it, yet they're not able to put it all together. We envision that we're back on the trail where we're approaching the bog bay. The chorus gets louder with each step until we're at the bog's edge, where it is so shrill that it hurts our ears. Every other Sound is blotted out: our footsteps, our breathing, the person ahead, the Owl calling in the distance—everything.

We continue on the trail, leaving the bog behind, and the intensity gradually subsides. With each step, the world around us comes more into focus as other Sounds join in the Frog chorus. Still, it is always there.

"Notice how you can tell exactly where they are," I state. "Their song has a sharp, piercing quality that makes no echo and is easy to key into. There's a reason for this, which I won't go into now. I'd like to leave it for you to discover. As you know, there's a reason for everything, so once you become the Frog and realize why you're singing, it'll probably become apparent as to why your song has this peculiar quality.

"Now, back to keeping on the trail. As you know from the

orienteering work we've done in the past, we need to know two things in order to get somewhere—where we're coming from and where we're going to. . . ."

"Aha!" says Kip. "The Frogs tell us where we're coming from, and when we come out of the Pines we're already pointed in the direction we want to go. So say we keep the Frog chorus to our backs, it should keep us headed straight for the road."

"It's that simple," I say. "Actually, it's even simpler than that. Once we awaken our aboriginal selves, we'll know intuitively what to do to stay on course. We'll extend our hands in trust and the mother of the night will take them and guide us."

13

Eagle Spirit
and the Tin Can

Preparing to Find the Invisible
Trails in Sky and Water

"You're telling me this is a canoe?" says my old friend Steve. "Looks more like a piece of scrap metal to me."

"Steve, how rude!" I gibe back playfully. "This is an Elder! Do you stick your mug in the face of other Elders and tell them what shape they're in? How about showing some respect?"

The boat—or what *used to be* a boat, according to Steve—lies upside down under some brush behind my cabin. It is late summer, the Wild Rice is ripe, and Steve has stopped over to see if he can borrow a canoe to go ricing. I don't take his reaction personally—it's just Steve. At the same time, I get the feeling he might not share my appreciation for the classic watercraft that I am allowing him the privilege of using.

He bends over the hull and scrutinizes her folds and dimples as though he were actually questioning her sea-worthiness. If he were a normal person, you'd think he'd stand back with an appreciative smile and savor the sight of such an ancient craft that has seen many waters and been part of many adventures. Okay, maybe his reaction is what you'd expect from a normal person, but I think I'll have some more fun with this anyway. I also wonder if Steve can see what I see in the canoe.

"What you're looking at are the marks of wisdom," I continue. "This boat has gained many teachings, many lessons on her journey. She wears her battle scars with pride."

"One teaching I don't need," says Steve, "is to be out in the middle of the rice bed in this old tin can with the water slowly creeping up over my ankles."

"Let me tell you the story of this boat," I continue. "Let's sit over here under the Pines, and then you can decide if you want to take her out ricing with you or not. But first, just take in her ambiance. She's telling us her story right now—all we have to do is listen."

"All right Tamarack, I'm usually up for a story, but it'll have to be a good one. And a short one, because the rice is ready and I'm itching to get out there."

"Yeah, it is," I reply, "and it's hanging heavy this year. After you hear this story, I have a hunch you'll just know whether ol' Bessie is intended to be your ricing companion or not. She could tell her story better than me, and you could probably hear a lot of it without my help. Those dings and dents you were looking at with a raised eyebrow are her voice. Right now she's telling you the story about when and where each one happened."

"'Story,' you say? It looks more like a nightmare. She must have gone down in Pearl Harbor."

"Hey, I thought you were going to listen! But you're right on one count—it is more than a story. It's a saga you'd hear if you could step back and keep quiet long enough to hear her song. There are all the voices, such as the age of the boat, the registration and park stickers, the repairs and how well they were done, and the color of the paint and how and why it's peeling. And then there are the scratches that show what kind of rocks she went over, the strength of the current, how much cargo she was carrying, and the ability of her paddlers. All these voices, and more that are not as obvious, spin the tale of where she's been and what she has seen."

"Hold up," says Steve. "You're not going to get me in that thing with a history lesson. Where's the story?"

"If you could give me a few minutes without interrupting," I huff back at him, "you'd find out. By listening to the song of the boat, you get to know her spirit, her needs and desires, her capabilities. She'll tell you whether or not she's a ricing boat, and whether or not she wants to go with you. In the same way that you have a need, so does she, and she speaks hers just as clearly as you did when you came and asked to borrow a boat. She sings a beautiful song, high and melodic. So let's give her our ear."

"Three minutes," replies Steve. "You can go heavy on the ketchup and onions, but hold the poetry."

It occurs to me that most good stories seem to have beginnings that are shrouded in the mists of time, and this one is no exception. This canoe sits here because of another canoe, and another canoe before that. This is the way of relationships that are part of the greater flow of our lives. A relationship with an individual may begin and end, and yet the flow of relationship is the flow of life itself. From the moment this old boat was born in some converted aircraft factory after the big war, I believe it was intended that she and I be together. When I was too young to know better and rode that block of ice down the rushing melt-water stream, this boat and I began our journey toward each other. A few winters later, when I made a boat out of a couple of inner tubes, some scrap lumber, and an old tablecloth for a sail, the distance between this canoe and me shrank a little more.

When I got my first full-time job, one of the first things I did was to buy myself a real canoe. I was too awe-struck by the fact that I could actually afford a canoe to mind the fact that it took some effort for two people to pick the thing up. Maybe that is why we had to pull her out from a tangle of brush when I bought her, and it is definitely why she sat behind my garage for a couple of years. I finally took the time to listen to her song, and she told me we didn't belong together.

In terms of cost per pound, there is no doubt that she was a deal. I would be the terror of a canoe demolition derby if there was such a thing. My guess is that she was probably one of the first fiberglass canoes and her builder didn't know when to quit. There was easily enough material in that boat to make three. No wonder she spent so many years as a hutch for the backyard Cottontails.

At the time, I didn't know anyone else who liked exploring overlooked wilderness streams to find colonies of Beaver who had never before seen humans. No one wanted to paddle out to a forgotten island to spend some time with animals who showed little fear because they were never shot at or chased by Dogs. I couldn't go myself, because I could hardly handle the boat alone. I know now that it wasn't my time to go out with others. I needed to be alone so that I could first grow in balance with my plant and animal Relations. I had much to learn, or I should say, unlearn. So rather than paddle, I swam and ran and climbed and crawled on my own.

I sold the battleship to a man with a young family. They wanted a solid, stable boat, and they had the energy to handle her, so they were happy and so was I. Somehow I knew that when the time was right, my intended boat and I would come together, so I didn't look for a replacement. I was content to wait.

About a dozen years later, I was living in a wigwam tucked back in the woods. One day I returned from a trip and was greeted by my neighbor, Larry, who lived in the little old farmstead out on the road about a mile away. He said some friends of mine had dropped something off for me while I was gone. Not many people knew how to find my lodge, so when out-of-area friends came by, they'd stop at Larry's place and he'd guide them out to me.

"What did they leave me?" I asked.

"I'm not sure," said Larry. "You might wanna come and see for yourself."

It was love at first sight. "This is her!" I exclaimed. "The boat I've been waiting for!"

"Excuse me?" replied Larry. "This, dear sir, is a pile of scrap metal. Where do you see a boat?"

"My sentiments exactly," interrupted Steve.

"Guess how I responded?"

"Ah, you told Larry he was being disrespectful of an Elder or some-such, right?"

"Not quite."

"Oh, ye of little vision," is what I said. "If we could use your car jack and a crowbar, pipe wrench, hammer . . . whatever you have, I bet we could restore this tin can to her old spunky self. You know, this boat and I have been walking separate paths, each waiting and preparing ourselves for this moment when we would be joined. I can see who she really is, who she wants to be again."

Larry whistled and shook his head, I think more out of pity for me than the boat.

My friends who dropped the boat off were die-hard canoeists who made regular trips to the Boundary Waters Canoe Area in northern Minnesota. It is a vast wilderness laced with lakes and rivers; in other words, a paddler's paradise. This boat had been with them on many trips over the years, and it looked like on this last trip they wrapped her around one too many rocks. In the shape she was in, I was surprised they even bothered to bring her back.

Enter the pity factor. They knew my financial situation. You need money to have a financial situation, right? They knew I liked to tinker, and they knew I hated to see anything end up in a landfill, so I figured that is where they got the inspiration to hand the crumpled mess over to me. Oh, and then there is the reason they probably didn't talk much about it—to save them the guilt of abandoning their trusty steed of many years.

But I didn't mind—I had a boat. In fact, I was grateful. And I was psyched. Larry went and grabbed his tools, which I took more as a sign of his friendship than agreement with me that the boat could be restored. That was okay, as I knew she'd float again.

"Ah, Tamarack," interrupts Steve, "didn't you say 'three minutes' about a half hour ago?"

"Just give me another minute—we're close to the end."

"Yeah, the end of the day."

"All right—I'll wrap it up quick. Promise!"

Larry and I worked all afternoon, first prying the gunwales apart so they were again on opposite sides of the boat. Then we used whatever tool worked best to coax the creases and dents into something that resembled the contours of a canoe. While we worked, we referred to her as a tin can—among other things—and wouldn't you know the name Tin Can stuck? To this day I still call her the Tin Can, and so does everyone else who knows her.

The Tin Can was made out of aluminum, which we found has a bothersome characteristic—it stretches. Even though we did quite a good job of bringing back the boat's original shape, a good share of her skin was still covered with little dings and ripples that were just impossible to smooth out. She looked like one of those droopy hound Dogs with too much hide on their frames.

But hey, I had no complaints. Obviously, I could see only her beauty. I looked at her dings in the same way I looked at my scars and fondly remembered the adventures that gave them to me. This boat and I were truly Birds of a feather.

The sun was already getting low in the sky when we finished, and we hadn't eaten yet. Even so, I asked Larry if he wanted to go on an inaugural paddle.

"Sure," he said without hesitation, so we grabbed some munchies, loaded the canoe, and took off.

At the time, I was scouting around for a new campsite. I wanted to move further back into the wilderness, and now that I had a boat, I wanted my camp to be accessible by water. I'd been exploring the head-waters region of a river about an hour away, and now with a boat, I was anxious to see what the area looked like from the stream.

"But it's late," protested Larry.

"Late for what?"

"Late in the day, it's gonna get dark!"

"And . . . ?"

After a moment's hesitation he shrugged and said, "Yeah, no big deal. It'll hardly be the first time we've been out after dark."

An old logging road took us within sight of the river, and we had about an hour of daylight left when we started paddling upstream into the setting sun. We silently swept by Beaver, Muskrats, Ducks, and Herons going about their business. The river narrowed and we had to pull ourselves over one downed Tree after another. At right about sunset, we came to the base of a canyon where the stream turned too fast-flowing and rocky for us to proceed any further by boat. We continued on foot.

Within a hundred paces we found ourselves standing on the massive roots buttressing a great White Pine growing right on the river bank. I felt drawn to look down into a depression between two roots, and there lay an Eagle feather—a large one from the middle of the female's tail. As you know, females are one and one-half times larger than their mates, so it is not too hard to sex a wing or tail feather.

"This is a good omen," I said to Larry. "When I see an Eagle, I consider myself blessed. To have one gift me a feather like this is even more meaningful."

I looked around and saw several smaller body feathers. Something else had to be going on. Sure enough, we looked up into the Tree and at the very top rested an immense platform Nest. It was a good area for my new camp.

A couple of days later, I came back and followed a Deer trail running along a cool, crystal-clear side stream to a rock-rimmed pool. Right there I built a sweat lodge, and on the bank above I put up a wigwam. My first full-fledged camp deep in the wilderness!

Every time I paddled up or down the river, I'd pass by the Eagles' Nest. We got to know each other. When the season turned and their fishing waters froze over, they'd leave for open water, and when the snows melted they'd return to raise their young. Occasionally I'd see one of them in the deep of the white season and wonder why they'd come around at that time of year and what they were eating.

While snowshoeing through an open Cedar swamp one crisp afternoon, I heard the chortling of contented Ravens up ahead. It sounded like they were feasting on a winter-killed Deer. The snow was deep and twigs as thick as my little finger were all the Deer could reach to eat. They filled their bellies, but wood fiber has no calories to keep an animal warm.

The sentinel Raven spotted me, gave an alarm squawk, and all five of them flew up into the treetops where they could see me. With them sat an Eagle, not a typical companion of Ravens. And it wasn't just any Eagle. I knew him by the way he stood solidly on his left leg while he felt around with his right foot to adjust his perching position. Sometimes he'd raise his right wing at the same time. I recognized his coloration as well; his chocolate body feathers were just a shade lighter than his mate's.

The area around my camp was so rich with lakes and streams that I could paddle just about anywhere I wanted to go, so I spent a lot of time on the water. More than once I was surprised to find the Eagles fishing on lakes several miles from their Nest, even when there were good fishing waters closer by.

And it wasn't just Eagles. I was continually discovering how different kinds of animals were more alike than not. A Bee went from flower to flower for nectar, taking a little here and a little there rather than sucking one blossom dry. A Deer kept moving while browsing, instead of completely stripping one bush. They were all providing themselves with a varied diet and balanced nutrition. And along with that, they were minimizing their impact, pruning, and pollinating to help assure their offspring would also have food.

Over time, I came to feel the spirit of Eagle at the Nest site, whether or not the Eagles were around. Their relationship to their Nest seemed to be much like mine to my lodge. They'd wander, sometimes spending time in different places, but they'd always come back to rejoin their mate and raise their young together. Their lives revolved around their Nest—it was the center of their universe—so there dwelt their spirit.

"Hey Tamarack, I hate to interrupt," says Steve. "This is a great story and all, but I don't see what it has to do with your Tin Can anymore. Or with ricing. I came over to get a canoe for ricing, remember?"

"Can you spare another couple seconds? I think you'll see real quick where I'm going with this."

"'Real quick,' yeah. Where have I heard that before?"

"I'll skip ahead a few years and wrap it up, okay?"

Steve rolls his eyes.

"Short and sweet, Steve," I assure him. "Here we go."

One spring day several turns of the seasons later, I passed under the Nest Tree and caught sight of a feather floating in a little pocket of quiet water. It was a wing covert—a medium-length feather from the shield area where wing and body meet. And it was an Eagle feather, but there was something different about it. Something oh-so-subtle that set it apart from the Eagle feathers I had found under this Nest in the past.

I took the feather out in the sunlight and studied it from different angles. I laid it down and took a couple steps back to gain perspective on it, and still I couldn't put a finger on what was unique about it. Maybe it was just the feel of the feather in my hand, or it could have been a tad shinier or narrower than what I was used to. I didn't feel a pressing need to pinpoint the difference—it felt good enough that I could tell there was a difference.

That afternoon I was upstream sitting beside a wide, shallow area that was the remnant of an old Beaver pond. I heard a call, then another. Two Eagles soared over the treetops and out over the adjacent marsh. They continued to call, seemingly disturbed that I was there. I figured they came hoping to do some fishing in the shallows.

Their calls . . . something about them didn't sound quite right. It seemed to be the pair I knew from the Nest down at the river, and the female's call confirmed that. Still, the male's was ever-so-slightly differ-ent from what I was used to hearing. Like the feather I found earlier. I wondered: did sickness, or poisoning, or maybe an injury affect the molt and cause a voice change? Could it be hormonal? Or the result of aging?

Over the next moon, I watched the pair from a distance, to see if I could find anything else out of the ordinary.

I deliberately kept from focusing or analyzing, so I could maintain perspective and be open to any nuance or impulse. I knew from experi-ence that as soon as I quit being as a question and tried to find a hard answer, I'd be stuck in my rational mind and not able to benefit from my senses, intuition, and ancestral memories.

I gradually became aware of something missing rather than some-thing new. The male didn't adjust to his perch with that quirky right-foot dance. He was a different male!

But did it matter in the scheme of things? There was still a pair, still the Nest, and it would remain the center of life. Every season, Fish would be caught and eaglets would be fledged. The age of the dead tree-top and the condition of a fallen chunk of the Nest showed a pair of Eagles may have called this Nest home for close to a hundred years. I hadn't thought to question how that could be. Now I envisioned the female growing old and barren and the male coming back to the Nest with another mate, or the male latching on to a Fish too big to lift from the water and drowning because he couldn't let go, and his widow then joining with a coming-of-age male looking for a nest. In these ways, individual Birds come and go, while the timeless, ageless pair, the spirit of Eagle, lives on. It was this spirit whose home was the Nest, and the individual Birds were merely the current manifestations of the spirit.

After coming down from the euphoric rush of discovery, I was left with a deep feeling of communion for those two Birds, and for all Eagles. I realized that to track Eagle through time is not to track indi-vidual Eagles, but the spirit of Eagle—the undying continuum of Eagle life. Like an aboriginal clan's hearth and a colony of Bees' hive, the Nest is the center of life for a family of Eagles. Without a Nest, a solo Eagle will just wander and die, leaving no trace, making no contribution to the spirit of Eagle.

When we read an animal's tracks in the mud or the tracks on a canoe's hull, we catch only a glimpse of her continuum, her spirit. When we attune to the song of her track, we come into communion with her continuum—we become spirit trackers.

I drift off into the silent world of my own thoughts, and it looks like Steve has done the same.

"Oh, there's one more thing I'd like to mention," I say after a bit. "I think my coming to see life and tracking from the greater perspective helped me understand why my Indian friends and Elders honored Eagle spirit in ceremony rather than the actual Eagle. It's the same for me and this canoe. Her spirit may have been with our Ancestors who paddled the waters of long ago to fish and gather rice, and her spirit could travel with our children's children on their paddling adventures. . . . Well, that's it, Steve. End of story. I guess I got a little carried away, eh?"

"Yep."

"So, do you wanna take the ol' Tin Can?"

"Does it come with a good lifejacket?"

14
Following the Flight Trail

And Now the Test:
Reading the Tracks Left by
Wings through the Air

"Phil and me, we were sitting out in front of the lean-to earlier," says Antoine in his Canadian bush accent, "and this Falcon flies over our heads and lands in the . . ."

"Wait," I interrupt as I unbutton my shirt to cool off from the run into camp. "If I may, I'll tell you what happened. She came in from the northeast, flew across the clearing, and landed over in that dead and leaning Fir Tree. And then she flew off toward Strawberry Meadow, flying low at oh, I'd say around shoulder height."

"How do you know?" says Phil. "You weren't there!"

"Are you sure?"

"Well, we didn't see you," replies Antoine. "Were you sneaking around?"

"Not this time," I reply.

"Then I suppose you know what kind of Falcon she was too, eh?"

"I might know what kind of *Bird* she was. Are you sure she was a Falcon?"

"I guess," continues Antoine. "She had pointed wings like one, ya' know? But maybe she wasn't. Don't jump to conclusions, right?"

"Can you describe her?"

"Yeah," says Phil. "What do you think, was she as tall as this?"

Phil holds up his forearm and points from wrist to elbow. Antoine nods.

"She had a wide breast," Antoine continues. "It was white, or maybe light tan, eh, Phil? And it was mottled with brown. Her back was brown too, eh? And she had a dark-striped tail."

Phil and Antoine are friends from the same region in Canada, which you can tell by their accent—and it only gets thicker when they're together. They work well as a team, with Antoine being upfront and Phil reflective. Phil had a troubled childhood and he is here looking for the meaning of life, while happy-go-lucky Antoine doesn't seem to have a care in the world.

I sit down with them around the lean-to hearth, ask for a piece of paper, and draw the flight silhouettes of our three common raptor families. There are buteos, the soaring Hawks, who have long and broad square-ended wings with short, fanned tails. And there are accipiters, the woodland Hawks, with their short, broad wings and long, narrow tails for maneuvering through the Trees. Falcons are the swift pursuers—they have narrow, pointed wings and long, very narrow tails.

"Our mind is designed to identify symbols,"* I explain while sketching. "We do it intuitively—we're programmed to relate to forms and shapes. This is true for most animals. A small Bird reacts not to Hawk himself, but to Hawk's silhouette. Have you ever seen a large window with the silhouette of a Hawk taped on it? People do it to prevent Birds from crashing into it. The instant a Bird sees a well-designed silhouette, she puts on the brakes and she is outta there. Unfortunately, most silhouettes are ineffective because they lack movement and shadow, which prey animals are very sensitive to."

If our hunter-gatherer Ancestors had published a field guide to Birds, it wouldn't be organized by family, genus, and species. It probably would have been a collection of silhouettes, along with the habits and patterns of the Birds who fit them. Aboriginal people didn't categorize Birds by genus and species. They went on the personal relationships they had with the Birds, as opposed to us, who operate on how someone else tells us they're related to each other. Even though Cranes and Herons have no close genetic relationship, the Ojibwe in my area see them as kin because they are of similar size and shape and live together in the wetlands.

That said, we still rely on a form of silhouette in our everyday lives, and in fact, we rely on it heavily. Silhouettes representing objects, thoughts, and feelings were the basis of our first forms of writing. With time, the silhouettes evolved and became more abstract, to the point

---

*For more on symbol tracking, see appendix 3.

where they were no longer meaningful in and of themselves. What they represented had to be learned and memorized in order to string them together to communicate.

Silhouettes finished, I hold them up and ask if they see their Bird.

They both identify her as an accipiter.

"Why so?" I ask, wanting to encourage them to note the differences in the symbols.

"It's the combination of long tail and short, rounded wings," replies Antoine. "Couldn't be anything else."

"So the wings on your Bird weren't pointed after all."

"I guess not," says Phil. "I just thought they were pointed because they were so short compared to soaring Hawks. Ah, looking at the outlines, soaring Hawks must be buteos, eh?"

"Would you like to name her?" I ask.

"Why? She already has a name," responds Antoine.

"I mean a personal name," I reply. "The one in the books usually doesn't say anything about her. If you named her, I bet it would reflect the Bird. And it would have some meaning for you. Another reason I'm suggesting you name her is that we place so much importance on learning the proper name that once we have it, we think we know something about the animal. It dampens our curiosity."

"Yeah, you're right," responds Antoine.

"I don't know that I'm right or wrong," I reply. "It's just a perspective—one of many."

"We'll let you know what we come up with," says Phil. "Right now I'm curious about something else. When the Hawk swooped in, the Birds on one side of the clearing raised a stink, and those on the other side went about their business as if that Hawk wasn't even there. It doesn't make sense to me."

"The Birds on the far side of the clearing," I state, "were the disturbed ones, and those in the Spruces down the hill from Hawk were the ones who stayed calm. Is that correct?"

"Ah, why do we even bother telling you our stories?" Antoine says with an exaggerated grumble.

"Even though that's a rhetorical question, I think I'll answer it anyway. To begin with, I don't know all the stories—far from it. And with those I do know, every time I hear them I learn something new. You bring me the gift of your eyes, which see differently from mine, so you'll notice some things that I miss. You perceive things differently from me, which helps me transcend my limitations. In these ways, I grow through your experience. I don't see how I could ever tire of hearing the same story again."

"I hear you," says Antoine as he feeds the fire to ward off the evening chill. "It's good to know we're not wasting your time. Now, can you give us some guidance on why those little, defenseless Birds just carried on like Hawk wasn't there?"

"This mystery I'm going to leave with you, because I think you know enough that before long you'll have it figured out. One clue: note when and where these Birds react to you in the same ways they did to Hawk. Remember, this is not about you and the Birds, it's about kinship with Hawk. When you start to feel that, the life-dance of Hawk will start to become your dance and it'll begin clicking as to why some Birds can ignore you—that is, *seem* to ignore you—while others in the flock are freaking out."

"Okay, I can wait," replies Antoine. "But something else about this Hawk thing puzzles me even more. If you weren't here, how did you know where she flew in from and that she landed on that certain branch and then flew off toward Strawberry Meadow? Did you see it happen before and just figure it was the same Bird doing the same thing out of habit?"

"Yes and no. I don't know that I've seen that particular Bird before, but I know the Bird's kind and I know she was a female. That is, as much as I know anything, of course. *Be as a question*, you know. And by the way, the branch she perched on was just an educated guess."

"But how could you possibly know species *and* sex?" asks Phil.

"Because I fly in the shadow of the Bird."

"Huh?"

"A particular Bird," I explain, "is just the present manifestation of her kind, and each individual flies on the track of her kind. When one Bird dies, another is born, picks up the track, and shadows the Bird before her. Usually to the same Tree, sometimes to the same branch. She then flies off in the same direction, even though the Bird before her may have done so days or even seasons before. So she's not literally following the Bird, but she's slipping into the Bird's lingering shadow, which is why it's called shadow tracking. Some call it spirit tracking, because there's no visible sign to go by."

"What you're describing sounds like a choreographed dance," says Phil as he reaches out to warm his hands. "Only the partners aren't together on the dance floor at the same time."

"That can be a helpful metaphor," I reply. "It illustrates why I sometimes call it a dance. And why the dance of this Hawk is the dance of all the female Hawks of her kind who follow in her footsteps. They all know the moves because they're imprinted in the genetic memory, and each Hawk can follow the dancer before her because they all have the innate ability to shadow track their predecessor."

"Then this isn't something a parent teaches her kid?" asks Antoine.

"Not really. The spirit of the Bird is shared by all of the Birds of her kind, so all of them share in the same spirit way. They have the same way of seeing and not seeing, the same way of acting and reacting, the same way of seeking shelter and raising their young. And they have the same way of flying into a clearing and choosing a perch with particular characteristics."

"That's hard to wrap my head around," says Antoine. "I thought Birds just flew around and kinda did whatever struck their fancy—you know, *free as a Bird,* eh? What you describe is, like, another reality. I understand what you're saying, but wow . . . I guess I've got to get used to it. Shadow tracking's pretty far-out stuff."

"You know, tracking is tracking," I respond, "whether it's you or a Bird doing it. Same skills, same principles. Bottom line, those Birds are following a trail. Whether you call it a shadow trail or a spirit trail doesn't matter—it's a plain old trail, just like any other. That Hawk followed her intuition, listened to the song of the track, and read sign the same as we do, the same as any other hunter does.

"Let me describe the shadow trail a little more; maybe that'll help. If you recognize a trail, you can follow it. I'll use the example of your Hawk's trail. It exists in two places: across the sky and within the Bird. These are actually more like separate dimensions, because they're so different from each other. Yet they fit together like a hand in a glove. First, the shadow trail exists across the sky. A sky trail is just as visible to a Bird as a ground trail is to us. A Bird can hear the song of the trail and follow it. She can read sign—she can 'see' the track of the Bird who flew before her. And second, the shadow trail exists within the Bird. It's instinctual, you might say. It's imprinted so clearly in her genetic memory that when the clearing first came to be, the first Hawk there knew exactly how to fly into it and which branch to light upon and which way to leave."

"I need another example," says Antoine. "How's about something other than Birds?"

"Sure. Imagine planting Fish in a lake that never had any. They'd know where to find food and which stream to swim up to spawn, even if none of their kind ever did it there before. And you'd know it too if you knew the dance of the Fish. You'd be able to call it, just like I did with your Hawk. There's no great mystery to it."

"That's good," says Antoine. "That explains why Hawk did what she did, but that still doesn't explain how you knew about it—how you knew her dance, I guess you'd say."

"How might *you* go about learning it?"

"Well . . . if I was one of those Hawks, I guess I wouldn't have any problem. Other than that, I'm lost."

"And that's just how I did it," I reply. "I became one of them. Time after time, I had watched them, but that wasn't enough. I was still separate from them—I still didn't know them.

"Once I could see through her eyes, I could shadow track her kind and learn their dance. Before long, I was Hawk."

Antoine and Phil lean forward for more.

"I felt her hunger and her fear and her need to protect her young, and it turned into my hunger and my fear and my need. I found out that her dance wasn't all that different from mine. It just existed in another realm, the sky world, which I didn't know. Once I crossed that boundary, we were kin. We knew each other; we flew and hunted together. Whether we had hair or feathers didn't matter anymore, because we shared in the dance.

"When I join in the dance of our Hawk, I come to know all the Hawks who've visited this meadow in the past, and I know the Hawks who'll come in the future. They are all one Bird, one spirit. It is only our minds that separate them from each other by time and space. In the same way, I am Deer and Mouse and Fern and Aspen when I enter this meadow. I track their shadows and dance their dance rather than mine. I am no longer me—I am merely what I am made of. If what sun and water and earth and fire gave me were taken away, I'd be nothing. Without venison and berries, without dreams and lessons learned, I'd be only a shell.

"When I recognize that I am not really me, I free myself to become what I am made of. I can be Mouse and I can be Hawk, because I already am them. Now I have no trouble flying. I have no trouble seeing how my Hawk sister chooses a flight path that keeps her inconspicuous so that she will not alarm the small Birds she hopes to feed her young. It is obvious as to why I have short wings and a long tail—they allow me to fly between branches and execute the tight maneuvers needed to catch swift, little Birds. From the vantage point of the perch I chose for just this reason, I can see distant openings in the canopy that tell me where there are other clearings. If I flew into a clearing from above, my silhouette could send everybody into a panic. When I swoop down low and silent, I can sometimes materialize as though I was always there.

"If I, Tamarack, become my inner Hawk, I don't have to climb the Fir Tree to see the worn spot on the branch where the she-Hawk always lands. I don't have to check for scat or feathers under the perch to confirm who the perch is used by. You no longer have to tell me the particulars of a story—or even the story, sometimes—for me to know what you saw. After your first few words, I could take over and tell the rest of the story, because it is me."

This is the secret of how natives hunt together silently. When they each know the dance of the animal they hunt, there is little need to communicate beyond staying in touch with one another. They each assume their role in the hunt, just like the Wolves of a pack. With Wolf and human, all that is usually needed to coordinate the hunt is a glance or slight twitch of the head.

Such subtle yet highly effective communication is made possible by being as a shadow, the state of consciousness of the native hunter. The modern person generally dwells in an ego state, which leads to mainly self-focused communication. It dominates the consciousness and numbs the individual to the subtle shadings of the shadow. From ego perspective, the shadow hunter can appear to be detached and unexpressive. Hence the stereotype of the stoic, aloof native. In actuality, he is no different from she-Hawk, whose demeanor is also subdued and straight-faced. At the same time, there is a flurry of passion and connection going on inside for both of them. *Stoic native* is not only an oxymoron but also an expression of one of the fundamental misunderstandings behind the prejudice toward native people.

If Phil and Antoine knew the dance of the Hawk, all they would have to do is glance up at the branch where Hawk alighted and I'd know what they witnessed. Having shared the experience in a moment's clarity, the three of us would probably burst into smiles of mutual awareness. To others, we'd appear silent, where in actuality we'd be communicating on the deepest level.

This can be hard to see for those who haven't experienced it themselves. We tend to form judgments based on our own sets of experience, which can throw us way off base and we end up inadvertently perpetuating false stereotypes.

We have talked into the night, the fire has burned low, and the two young trackers have grown silent. The days are full and rich here in the wilderness, and it is time for them to turn in.

As passionate as they are to awaken, I fear that I sometimes overwhelm them. It is a constant balancing act for me to give them tantalizing tidbits and stop before they get more than they can digest. When this happens, they either have to reject it or take it on faith. The first option is healthy—the second, dangerous. In the natural realm, faith is fatal. If it is not real to you—if you haven't seen or experienced it yourself—it might be best to stay with what *you* know. Perhaps the day will come when it enters your realm of experience, but until then, let it be.

I leave Phil and Antoine with these parting words: "Keep in mind that what I share with you is not gospel. It's only what I've come to know, in the way I've come to know it. I miss more than I catch, and there are things I skew because I don't catch them clearly. Someone else might describe what we talked about tonight in another way. This is natural, as we each have different eyes and different experiences to draw from. So hear what resonates, let go of what doesn't, then spread your wings and join in the dance."

15
Human Tracking, Bear Style

Bear, the Master Man-Tracking
Teacher, Shows How to
Find and Trail People

Advertisers and inspirational speakers have beaten the slogan "challenge is opportunity" to death, but in the wilder places it is alive and well. In fact, you could call it the wilderness creed. Everyone there lives by it, and newcomers either adopt it or fade away. This may sound cruel and oppressive to those not familiar with nature's way, but those who do know it find that embracing the creed transforms fear into curiosity and scarcity into abundance.

Mike has just stepped off the plane. Being from the Northwest, he feels right at home on this cool, drizzly morning. It is the first day of the yearlong course and right away Mike has a chance to find opportunity in challenge. He, along with nine strangers just as green as he is, are preparing for a soggy six-mile hike over rugged terrain. Their destination is the primitive camp of bark and thatch lodges where they'll be living for the next turn of the seasons. Rick, my assistant, will guide them, and only he knows that I'll be trailing the group.

After about a mile, the trail sign tells me the group has broken up. I'm out of sight and hearing range, so I don't know all the particulars. I see from the sign that one woman—it looks like Mia—lagged behind. A few at the rear of the line slowed up also, probably to keep her company. My hunch is that Rick considered this a good time for everybody stop for a rest. Raven and Jay are telling me that they left from there in two groups, probably to accommodate different rates of travel. The next two miles skirt a lake and follow a stream, so there is little chance of anyone losing their way.

Mike heads off on his own. It doesn't surprise me considering his independent streak, which became obvious shortly after he arrived by his impeccably organized gear and apparent aloofness. I knew better, however, as I had gotten to know him last year during his weeklong visit to the program.

Right now he doesn't know I'm shadow tracking him. I stop to watch Raven quietly approach her nest to feed her young, as it appears Mike did, and I catch sight of Woodpecker working over a dead Aspen Tree off to the left. I wonder if Mike saw her also.

Knowing he was unsure of where he was going, I keep an eye out for a spot where he'd likely have stopped to orient himself. Surrounded by heavy growth, the low ridge he came upon wasn't the ideal place to catch sight of what lay ahead. Still, it was the best view available. My hunch is that he went for it. I cut over to the spot he was most likely attracted to, which was a tiny Hazelnut-choked opening nestled between young Firs.

I can feel his presence right here on the cusp of the ridge. It doesn't matter whether he is here with me or whether he was here a short while ago or even the day before. His shadow lingers and I know its feel, because I've been walking in it now for around a mile.

When listening to the song of the track—and especially when following it—I find that holding on to any concept of time just gets in the way. I must be completely in the now to hear the whole song, because it's being sung in the present, echoed from the past, and reverberating into the future. If I were not in the now, I could look at a star and say, "It might no longer exist. The light that I'm now seeing took millions of years to reach me, and the star could have burned out in the meantime. If I were to track the star, I might find only blank space."

None of that matters when I'm in the now. The star does exist because it is part of my present reality. And it is in this state of presence that I can best track. Any distance that I create between the tracked and me by thinking about how separated we are by time, such as how long ago he passed by this spot, muffles my ability to hear the song of the track. I run the risk of reverting from my heart-of-hearts to my head and tracking analytically. When I need to check in on the time elapsed in order to get a sense of where my quarry might now be, I do so quickly and get right back to the present. In this case, I want to be Mike, right here and now, thinking and feeling and doing just as he would. I couldn't do it if I kept track of the passing time, as time is a mental construct that separates me from the reality of the now.

With the haze and drizzle limiting visibility, I feel his anxiousness. Every direction looks the same. I nervously reach over to snap off a dead twig for something to fidget with.

Right where I reach, a twig is already snapped off. It is torn upward, the way it would be if a human broke it off. It's a side branch rather than the tip, which is what would most likely be broken off by a passing animal. It is at the right height for Mike to have done it. And the break is still dry, indicating it happened very recently.

The route he is taking is predictable, so even with his momentary uneasiness, he is in no hurry. He exudes confidence, and he knows he can take most of the day for the trek. I circle ahead to find a place or two where I can become invisible beside the trail I believe he'll be taking. I'll be spending the next year with him, guiding his awakening along with being responsible for his welfare, so I want to learn all I can about him. The way he is when he is alone in the woods is far more telling to me than how he is with others.

If ever Mike were to get lost and I had to track him down, odds are he'd be by himself. Confident or not, someone who is lost is typically in greater danger when alone than with companions. If there is any secret to lost-proofing, it is *don't try to think your way out*. The way out is in the sign, not the mind. Emotions cloud the mind, while sign is clear, objective, and reliable. Those who maintain composure, read the sign—and trust in it—can usually take care of themselves. Unfortunately, most lost people turn to their minds, which become liabilities by undermining perspective, even before panic sets in. People panic much more often when they're alone than when they're with someone, because they don't have the comfort of a touchstone.

I see movement ahead: something rocking back and forth. His wobbling gait gives him away before I can actually make him out as he emerges from the woods into a clearing. He goes out into the middle of the clearing to pee, where I suspect he feels safe and secure because he can see all around him.

What he doesn't seem to realize is that he can only see what is in the clearing. It is hard to see into the dark from a lighted area, so what lurks in the woodland shadows is hard for him to detect. I'm standing in full view about thirty paces from him and he can't make me out. It is much easier to see from dark to light, as in a movie theater. Look at how easy it is to peer into a house window at night and see what is going on. At the same time, someone inside could have a hard time seeing you.

Mike is only reflecting his urban upbringing, where he got used to the outdoors being lit by streetlights and yard lights. The popular wisdom is that night lighting helps ensure safety. The odd thing is that it does the reverse, by making shadows where unscrupulous people can easily lurk. At the same time, the light makes it harder for someone to see them. Even though the lights may be of some help, they create a clear-cut double hazard.

What are Mike's options? If he were a native, whether two-legged or four-legged, he'd likely hang out in the edge, where clearing and forest meet. Just inside the tree line, he wouldn't be exposed, and from there he could see both into the shadowy forest and the bright clearing. Right now, he sees what is going on in the clearing all right, but he doesn't notice me just a couple of steps into the woods, even though I've made no effort to conceal myself.

We seldom need camouflage to become invisible when we know the ways of the one we track. I slip into blind spots and use movement patterns, assumptions, and distractions as my disguise, as I did with Mike. This type of camouflage is readily available, and I especially like it because it helps me become the one I track.

On water as well, natives tend to stick to the edge. When paddling just off-shore, you're far less conspicuous to anyone around the lake than when you're out on open water. And like inland edge, shoreline and adjacent shallows are where the majority of life is found. With an offshore wind, shoreline waters can be calm even when the rest of the lake is choppy. The Ojibwe of my area use this chant to keep themselves paddling in rhythm: *jeemanakaganakeego; jeemanakagananodin.* It means *paddle your canoe close to the shore; paddle your canoe out of the wind.*

Again I circle ahead, this time choosing an observation site at a narrow spot in a dense Alder swamp where I think Mike will cross. There is an old rotted-down Beaver dam spanning the narrows that's just high enough to keep his feet dry. About ten paces off to the side of the dam is a tiny grove of Cedar with dense overhanging branches. Soft amber needles carpet the mossy ground, which invites me to sit down, lean back against a Tree, and take a nap. It is a quiet day with excellent sound-carrying qualities—high humidity, still air, low cloud cover—so I'm reasonably sure his approach will wake me up.

When he passes by, I feel like a gnome sitting under a Tree watching the activities of a world that doesn't know I exist. Yet he looks my way—directly at me, in fact—as though he senses something and wants to know what it is.

Even though he doesn't see me, I consider myself exposed. If I was completely in the now and in shadow mode, he shouldn't have picked up on anything. Do I consider myself a failure? On the contrary. I'm always in training, and I am thankful for this new opportunity to learn how to better get out of the way of my inner tracker.

For the rest of the hike, I fall back well behind Mike, where I can anticipate his moves before looking for sign to confirm them.

As I weave through an Alder thicket, I realize that my pockets are filling. Hey, what is a tracking adventure, at least for me, without mementos? This time it is a Beaver skull, the lower jaw of someone's long-lost hunting Dog, a pocket full of cranberries, and a couple of feathers that have their own tracking story to tell. The animal parts are from long-ago disturbed kill sites, or I'd have let them be. These tokens are props for the stories I bring back, but let's save them for another time. Right now I'd rather tell you about the tracking teacher Mike is going to have for the next year. No, it is not me. I'm doing well if I can stay out of the way and get someone connected to a real teacher. Finding that person—a true-grit tracker who lives by his trade—is a tough assignment in this day. There are so very few places left where trackers can practice their traditional ways.

One such place, the Kalahari of South Africa, is still graced with the presence of a few traditional !Kung San who can identify the tracks of everyone in their camp. As if that weren't enough, these masters can tell from the tracks how their campmates are feeling and what they're up to. At least outside observers claim this information comes entirely from the track. Unless these onlookers are aware of the song of the track, it is not surprising that they'd attribute what they see to what they know.

At any rate, as amazing a display of tracking virtuosity as that might be, the !Kung San have it easy tracking their own people. And I had it easy tracking Mike, one of my own people. Individual cases like these are the exception. Humans, in my experience, have become some of the most difficult animals to track.

The profession of tracking humans is commonly called *man-tracking*. It is not a term I like using, not only because it can be construed as sexist, but because it is misleading. Men, women, children, and older people each leave tracks specific to them, and each requires a unique tracking approach. Women, for example, tend to rotate on the ball of the foot, while men often rotate on the heel or toe. Look at the wear patterns of shoes and you should be able to see it. Still, this is not entirely true, as it applies mainly to urbanized men. If native men rotate at all, it tends to be on the balls of their feet.

I find it necessary to separate tracking native people from tracking urbanized people. Although there is overlap, two separate skill sets are needed. Not only do the tracks differ, but the movement patterns do as well. Each kind of creature is predictable in its own way, as are all humans. And still many who have experience tracking both native and urbanized humans will tell you it is like tracking two different species. Learning to track one helps with tracking the other about as much as studying a captive Wolf's feeding pattern helps with knowing that of a wild Wolf.

As if that doesn't make human tracking confusing enough, there are no clear lines between male and female tracks and those of native and urban people. There are modern people with native-walking tendencies, and vice versa. Some males and females have similar character types and physiques, which results in similar track patterns.

Contrary to tracking adults, tracking children can be sometimes refreshingly easier—but sometimes infinitely more confounding. The younger they are, the harder it is to distinguish their gender and the more urban children resemble natives in their movements. Under age six, it is hard to tell them apart by their trails. This includes both their walking style and their movement patterns. It is not until hormones, shoe wearing, and lifestyle differences start changing their physiologies that differences become apparent. I remember how surprised—and sometimes downright frustrated—I became as a child when these changes first started to show in my playmates. Where we once all dove into our woods and swamps adventures together, we started to segregate ourselves by gender and how down-and-dirty we wanted to get.

When talking about human tracking, I stretch the definition of native to include those who relate to natural environments like a native. This includes some urban people who have spent time in the woods. I don't mean time gauged just by the clock or calendar, and I don't mean time camping or hiking or hunting. That kind of time doesn't do much to change track patterns when people bring the city with them into the woods. I'm talking about attuned time, when time becomes timelessness and forest and self become one. After tracking someone for a quarter mile or so, I can tell which kind of time they spent.

They say Dog is descended from Wolf, but if their paw prints didn't resemble each other's, you'd hardly know from tracking them that they were related. If I were good at tracking Dogs and tried tracking Wolves, I might think Wolves were the most elusive of creatures. This has little to do with the cleverness of Wolf, but more with the inexperience of the tracker.

The same is true of urbanized and native humans. A city person who can track his own kind, say a search-and-rescue tracker, might hit a wall when trying to track a native. This is what makes humans one of the most predictable of creatures and one of the easiest to track, and at the same time one of the most confounding.

Here is where that tracking teacher of Mike's that I mentioned earlier comes in. This guy is the best teacher of human tracking I've found. He takes the difficulty out of it by helping us understand why we do what we do. But there are a few catches: he doesn't speak English well, he's temperamental, and you've got to catch him at the right time. If you can find him. And on a whim he'll hold a class in the middle of a swamp or garbage dump.

This guy's been my teacher for a while, by the way, and I don't mind his quirks. Everything inside of me said to go and learn from the best, so that's what I did, and I hope to set Mike up with the same teacher. This guy is Bear. No, it's not a nickname and I'm not pulling your leg. I'm talking about the four-legged furry Bear—he'll get you tracking humans like no human can. Need proof? Ask a Bear tracker and he'll likely confirm that Bear trackers make great human trackers, and vice-versa. This is because Bear is our closest relative.

I know, this conflicts with what you've been taught about us being most directly related to apes. Here's why the Ojibwe of my area, along with native people virtually everywhere who live with Bear, call him "Brother." Like us, he's an omnivore. He eats fruit, nuts, tender greens, Insects, meat—all the things we would naturally eat. We turn to the same medicines as Bear when we're sick. We can learn about new foods, medicines, and foraging methods by watching Bear. If we're starving, we can follow him and he'll feed us. A sloppy eater, he usually leaves enough to scavenge. He can then feed us again, by becoming our food. When we follow him, we learn his habits and patterns, which will help us in hunting him. And he's not hard to follow, as we both travel the same trails at around the same speed. Unless it's a race, then Bear wins.

Let me clarify a couple of things here before we go on. First of all, I'm talking Black Bears, not Grizzlies, and I'll tell you why. Grizzly Bears are nearsighted, so they can be surprised by someone walking up on them. And they don't take kindly to it. However, they hear very well, so people in Grizzly country have taken to wearing little bells to let the Bears know they're coming. This turns out to be a big help in telling whether or not there are Grizzlies around. All you have to do is take a look at the Bear scat, which is easy to find. If it has fruit seeds and hair in it, it's Black Bear, and if it has little bells in it. . . .

I'm serious. Grizzly Bears are normally quiet creatures who keep to their own business. At the same time, they're the most powerful land-based predator on the continent and they have no reason to fear any other animal. If a Grizzly gets riled for any reason, you don't want to be there. Even though Black Bears kill just as many people as Grizzlies, with twenty-five deaths attributed to each over the past twenty years, only a relative handful of humans are exposed to Grizzlies in their remote wilderness range.

The disparity probably has more to do with temperament than size. As apex predators, Grizzlies have little to fear. Although they generally run bigger than Black Bears, Grizzlies in central Alaska average only around 500 pounds, whereas Black Bears in my area range up to 700.

We have a third member of the family here in North America—Polar Bear. Living up to her name, her range is circumpolar, so anyone in the world who hikes far enough north could run into her. She can be just as dangerous as Grizzly, so you'll want to know how to identify her. If a Bear chases you up a Tree and comes up after you, he's a Black Bear. If he knocks the Tree down to get you, he's a Grizzly. If there's no Tree to climb, he's a Polar.

The second thing I'd like to explain is that I realize most of you won't be able to actually shadow track Bear, because he doesn't live nearby. That, however, is no reason for you to feel left out while I describe how to do it. You can still have Bear as your teacher by visiting places where he lives and following his trails. That alone can be an awakening experience. Watching videos of Bear in his natural habitat is a good way to get a feel for how he moves. And then there's learning by doing—you can become Bear and live his ways first-hand. Envisioning works well for a lot of people because it can be done anywhere. If you can imagine or fantasize, you can envision. I believe we evolved the ability to facilitate the hunt, and now in our modern lives we call it fantasizing or imagination. The bottom line is that we have such a resonance with Bear that just about anything you choose to do is going to bring results.

Since Bears and humans have similar movement patterns, following a Bear is pretty easy for a native. Most of the rest of us are going to struggle at first until we loosen up a bit. Once we're able to get a start, it should come easier and easier on its own. It literally becomes a no-brainer. The degree of ease is our barometer for how close we are coming to our natural way of moving. When it becomes second nature, we know we've returned to our natural state.

This ease of movement is the way of the native tracker. When he becomes the animal he hunts, hunting is no longer work, no longer the struggle of the hunter versus the hunted. Like a healthy, caring relationship with a loved one, it becomes effortless, enriching, and enjoyable.

Some of the students I work with are surprised when their ancestral memories and intuitive abilities, together with their tracking ability in general, improve right along with their ability to move like a natural-living human. There appears to be some intrinsic connection between natural movement and our inner tracker.

This is my main argument for encouraging aspiring human trackers, even if they'll never track natives or Bears, to start by learning to track them. Once they understand the norm, they can better understand divergence from the norm—in this case, the movement patterns of urbanized humans.

Whether you're into wilderness survival, hunting, trapping, or berry picking, you can become better at it by following Bear around for a while. He'll not only show you how to find the best areas, but he'll take you there. Again, you don't have to literally shadow Bear; taking his trails will do just about as well. But make sure it's a Bear trail. City people tend to follow Deer trails, and if you know what they eat, you know that not much of it is fit for human consumption. Deer's diet has so little fat and protein—two vital fuels for the human machine—that it wouldn't sustain us for long.

Young urban children, on the other hand, tend to follow Bear trails, along with moving through natural landscapes like Bear. The older they become, the more they choose high, open land over lowland with cover. They choose dry land over wetland. The trails they'll come across are mainly those of ungulates: Deer, Elk, Goats, and so on. I adjust my tracking style to fit the child, thinking like Bear to find lost young children and thinking like an urban dweller to find older children.

There is another reason the human-Bear connection doesn't click right away for some people—Hollywood. From Tarzan swinging through the Trees to screaming braves on horseback, we're given an image that is anything but casual and deliberate. The reality is that many native warrior societies follow sayings such as *train like a*

Cat, move like a Bear. Cats typically jump up on boulders and fallen logs in their path, while Bears go under and around. Even though a native might be perfectly capable of moving like a Cat—i.e., the way Hollywood likes to depict him—he usually does so only when necessary. When he's in stealth mode or stalking, he wants to stay as inconspicuous as possible, which means staying low and moving like Bear. Additionally, a warrior wouldn't risk twisting an ankle unless it was truly necessary—his people rely upon him for so many things.

We intuitively know this, because it's the way we evolved. When we're centered in our heart-of-hearts and hearing the song of the track, we naturally move and track like Bear.

When I want to track intuitively, I'll often become Bear. I follow the contours of the land and stick to the edge of the lowlands. I have no regard for direction; I let instinct and the landscape tell me where to go and what to avoid. When tracking a modern person, I switch to ego mode and draw the line between myself and nature. I don't trust what I feel and see. I stick to highland if I can, I keep track of time and direction, and I think a lot about what I'm doing and where I'm going. I worry about my performance, and I sometimes feel uncomfortable and out of place. I don't like bending over and going around things; I stay upright and push through obstacles.

I also become Bear when I want to avoid being tracked. Some people find it hard to believe I go where I do. They can only see me getting cold, wet, dirty, and lost. This allows me to go places—sometimes right under their noses—where it's likely no human has set foot since the natives were the only ones around. Those are the places where I can hide and no one would think to look. Some of them are little overlooked pockets of wild with families of animals that people wouldn't think lived so close by.

By now some of you must be wondering just how to become Bear. If you've been listening, it's probably already happening. I'd like to help it along with a little history lesson. Don't worry, it is natural history. Bear has very good close-up vision, while his distance vision is only fair. It's probably not even as good as yours. On the other hand, he can hear and smell much better than us. *Much* better—he outshines the best scent-tracking Dog several times over.

Along with moving at a speed that makes it easy for us to walk along with him, he quickly gets accustomed to having other Relations around, including us. He can be very generous at kill sites, allowing Vulture and others to eat right beside him. These qualities allow a human to be constantly in his range of sight and he'll go about his daily business as though nobody was there.

Plan on a lot of time for meditation, knitting, or whatever, as Bear likes to nap. I'm not talking just an afternoon nap here, but several naps spread throughout the day. The current record holder is a Bear who a friend of mine watched take eighteen naps in one day! This is even more impressive when you consider that Bear is normally a daytime animal and sleeps at night. At the same time, he likes his peace, so if there's human disturbance around, he'll become nocturnal.

A Black Bear with small cubs seldom wanders far from a large, rough-barked Tree. The furrows in the bark make the Tree safe and easy for cubs to climb. In case of danger, the Tree is the cubs' refuge, and the Tree serves as a napping roost. Considering how much Bears nap, you can imagine how important it is to have such a Tree nearby.

There is some danger in following Bear around, though it's not what most people think. And it's minimal if you know Bear. Keep in mind that Bear is a big and powerful predator, and that on occasion humans are preyed upon.* Before striking out on your own, get training from someone who is habituated to Bear or tracks Bear professionally. Mike and his classmates were thoroughly briefed on Bear country etiquette before they did any tracking.

---

*For information on Bear safety, see appendix 4.

When shadowing or becoming Bear, remember that he does what he does because of his design and his needs. Merely imitating his movements won't give you a feel for that; you need to see the world through his eyes and feel his hunger. Get down on his level, both physically and emotionally. Travel his way, forage his way.

There is one more consideration, and I saved it for last because here's where our personal stories come in. I watch some people study Bear ecology and tracks, all the time keeping a respectful distance. Why? Fear. When we fear Bear, fear becomes the basis of our relationship with him. To accommodate our fear, we adapt and adjust our behavior around him. We see everything through a lens of fear; it warps our perception of his habits, movements, and motivations. Imagine what potential your relationship with your best friend or lover would have if it was based on fear.

As if that's not enough, our fear blinds us to Bear's fears. If we knew what he cherished and disdained, what gave him the assurance and made him nervous, we could see the individual behind the image. Just like us, each Bear has a unique personality. If we could read his thoughts and foresee his movements, we'd have the potential for an interactive relationship with him rather than a reactive one. We could literally walk among Bears and be safe. (This is in theory, mind you. Don't even think about trying it unless you know Bear as well as your mother.)

Fear, as Mike will learn in his course experience, is only a lack of knowing. He might fear the sounds of the night until he finds out they're only the scurryings of Mice and the callings of Owls. There is a saying, *to know you is to love you*. Knowing Bear can help us love him for who he is. By becoming Bear, we can then come to know and love ourselves for who we are.

I'm the proud descendant of a lineage of hunter-gatherer trackers on both sides of my family. All of us, in fact, have solid hunter-gatherer ancestry—for all of human history until the recent dawn of agriculture, there was no other way to live. The gift I cherish most from Bear is that along with teaching me to track my own kind, he has shown me the track to my Ancestors. They walked with Bear, and I see myself carrying on this hallowed and timeless tradition, so that it may live on to greet my children's children.

16
The Messenger

A Frontier Opens:
The Tracker Attains Relationship
with an Elusive Predator

"I just talked with two women who saw a Cougar," exclaims Lety as we meet at our camp's trailhead. She has my rapt attention.

Mystical Cougar, the silent stalker whose ghostly presence is the symbol of some of the most remote and pristine areas left on the continent. It's been a century since the last surviving Cougar in the Great Lakes region was shot, and I'm always right there trying to verify any reports of sightings I hear about in my area. Whether they are tracks, hair samples, or pictures, they inevitably turn out to be Bobcats, Yellow Labs, or mangy Coyotes—even Housecats, if you can believe it.

At the same time, the training I've received from Ojibwe Elders has taught me to consider all the possible options, and to seek the deeper truth behind what I see or hear from others. On this cool, drizzly day late in the Blackberry Moon, I'd love to believe that those women have actually seen a Cougar just a quarter-mile down the dirt road from where we are standing. Only four Winters ago, the road gave us the first tracks to confirm that Wolves had returned after being gone for decades.

"What was the story the women told you?" I ask Lety.

"They pulled up at the boat landing," she replies, "where I was picking berries just before I came down here to meet you. They said they were driving down the road looking for Mushrooms, and just a little ways back this big, gold-colored Cat with a long tail stepped out in front of them."

"Did you ask if she might have been a Dog or a Coyote?"

"They swore they saw a Cougar," Lety replies. "They said she was a lot bigger than a Dog, and that she was long and lanky with a small Cat-like head."

Sounds like a Cougar. At the same time, I recall how my imagination created a Muskie out of a big Sucker Fish I caught when I was a kid and didn't know the difference. I was taught that Muskie was the King of Fish and Sucker was a lowly bottom feeder, so I wanted my first big Fish to be royalty. The power of suggestion works similarly—when

someone in the neighborhood reports seeing a prowler, others start sus-
pecting innocent strangers.

"Did you get their names and addresses?" I ask. "I'd like to get the
exact location from them, and I'd like to quiz them more to see if it
might've been another animal."

"Ah, I wish I had," groans Lety. "I remember they said they lived on
Wigwam Lake and the younger woman was . . . ah . . . Laura I think.
She looked like she was in her fifties. The other one was Madeline. She
had a strong accent, probably seventy or eighty years old."

"How about their license plate number? We could track them down
that way."

"No, but I remember the vehicle. It stood out—it was one of those
little Jeep-type vehicles, royal blue."

"That should be enough to find them," I reply. "There aren't many
cottages on Wigwam Lake."

"Don't we want to look for the track first," suggests Lety, "in case it
rains again or it gets driven over?"

"Good idea. They're more likely to be home later anyway."

An early morning rain has softened the sand and the overcast sky
has kept it damp—perfect for holding a track. On top of that, so few
vehicles come down the road that tracks sometimes hold for days.

Now if we found a Cougar track, what would it prove? Only that
we found a Cougar track, I'm afraid. The official positions of the
Department of Natural Resources and Forest Service are that no wild
Cougar population exists in Wisconsin, and that any confirmed sign is
from pets who have either escaped or been released into the wilds. In
many cases, I believe their pet theory is valid.

But does it matter? Anyone who's let their inside-raised Cat out
to roam knows that it's only a matter of days before he's pouncing on
Mice and swatting Birds out of the air with the best of them. Even if
declawed, they're going to stalk and kill. Housecat or Cougar, I don't
think it matters. Sure, they differ in size, but they have one fundamental

trait in common: a passion for the hunt. It so defines Cats that virtually everything they do is somehow related to the hunt. It shows right away in kittens, as most of you have seen. As soon as they can walk, they start pouncing on one another and attacking mom's twitching tail. Even the shyest Housecat will paw, bat, chase, or wrestle with something. It's all in training for—you've got it—the hunt. I'll bet that if Cats dream at all, it's about hunting.

The Cat family evolved as pure carnivores. They have no molars for chewing, just specialized teeth for grasping and slicing meat. Cats' only divergence from pure protein is a few blades of Grass now and then to scour their guts. And it's not just flesh they prefer, but live flesh. Unlike their canine cousins, they don't normally go for carrion, and unlike most canines, they'll hunt even when well fed. It's their nature; what else can they do?

This makes it possible for Cougars, even if caged from birth, to be released as adults and not only survive but naturalize themselves. Some wildlife managers hold the position that released Cougars cannot form the nucleus of a wild breeding population. On this one, I have to go along with what Cougar tells me.

One reason Cougar can rewild herself so easily is that she's an anomaly. Specialization, as with Cougar's diet, is generally a limiting factor in the natural realm, yet Cougar has the most extensive range of any large animal in the Americas. She's found from coast to coast and from northern British Columbia to Tierra del Fuego. One reason is that meat is highly nutritious, rich in minerals, and easy to digest, so she needs to kill and eat only occasionally. Herbivores, who subsist on nutritionally poor and hard-to-digest plants, need to spend a good share of their time eating.

I never knew Cougar back when she was a regular Northwoods resident, but I feel her absence. She had been here for eons, right along with Moose, Wolf, and Deer. When a community loses a member—especially an apex predator like Cougar—the whole Hoop of Life is affected. Cougar is designed to hunt and eat Deer, and she'll live on Deer almost exclusively if they are available. Deer need Cougar just as much, to keep them healthy by weeding out their sick, old, and over-abundant young.

The entire forest ecosystem benefits from this circular relationship. Plants don't get overgrazed, scavengers from Raven to Bear have carcasses to scour, and there's enough forage to go around for other browsers such as Moose and Elk. Grouse and Beaver gain when young Aspen, their favorite white season food source, isn't wiped out by a horde of starving Deer. The chilling scream of Cougar is as much a part of the Northwoods as are Spring Peeper's shrill chorus and Barred Owl's ghostly moan. Cougar didn't leave by choice, and still maybe she'll choose to come back. Her woodland Relations, including me, miss her dearly. As soon as it is clear to me that she has returned, I'll hold a welcoming feast in her honor.

Some people think that with Wolf here as Deer's guardian, we don't need Cougar. True, both Wolf and Cougar inhabit the same ecological niche, and still in many ways they are as different from each other as night and day. Cougar is a solo hunter who likes to pounce from above, while Wolf hunts as a pack and relishes the chase. How does she do it alone? With tremendous athletic ability. Wolf, with some effort, can pull himself over a barrier as high as a tall person can reach. You'd have to double that height to challenge Cougar, and she can do even better on a horizontal leap. This is from an animal no bigger than you or me. Females run from 90 to 120 pounds and males are about half again as large. How, then, does she do it? No problem when you have the largest hindquarters for your size of any Cat.

The male cougar has another advantage in his species' struggle to survive as lone animals: a muscular penis that allows him to spray the undersides of leaves and branches. In this way his scent is protected from the elements. This is just as important as his gymnastic skills, as it helps him effectively mark and maintain both his far-ranging hunting territory and the females' territories within his.

The paths of Wolf and Cougar seldom meet, as Wolf is a highly visible pack animal and Cougar travels alone, slowly, and secretly. Cougar's keen senses and quiet ways keep her in the shadows and out of trouble. She's the phantom of the forest, able to live for years in an area before any human comes to realize it, if at all. A female's territory can be small, extending only as far as needed to find Deer. A Wolf pack, on the other hand, needs to range far in order to feed all its members. To assure an adequate food base, the pack maintains its territory by regularly patrolling and marking boundaries. This is no problem for Wolf, whose loping gate can carry him all day without tiring.

Herein lies a fundamental difference in hunting styles—Wolf actively seeks his prey, while Cougar lets it come to her. At the same time, there is a fundamental similarity, as both Wolf and Cougar must employ their wits in order to position themselves in the right place at the right time.

Along with the difference in distance covered, they each move differently. Slinky and sinewy Cougar flows over the landscape like water. She's not a runner; she pads gingerly along, carefully picking each step. Her softly cushioned feet leave a track visible only to the experienced tracker. Wolf is built stiff and springy, more like a taut bow. When running, he flies over the ground like an arrow, with his clawed feet often leaving easy-to-find tracks.

So that's the story as to why solid photographic evidence of Cougar's presence is hard to come by. Most purported Cougar photos show a distant, out-of-focus animal blurred by shadows. It is in the white season that her presence can be most easily detected, by her tracks in the snow. However—true to her nature—there is another glitch: her tracks are few and far between because of her sedentary hunting style and slow-moving ways.

There were once three major obstacles to Cougar's return: suitable habitat, a food base, and our fear.* When the Northwoods was leveled by loggers, market hunters and poachers virtually eliminated Deer, Moose, Elk, and Caribou, and bounty hunters got what predators the loggers and settlers missed. Now the forests have grown back and the Deer population has rebounded to levels the Northwoods has never seen. Only the negative attitude toward predators persists, like a festering wound that refuses to heal. Even some of us who welcomed Wolf's return shudder at the thought of a large Cat slinking silently through the woods and lunging from an overhanging branch onto whatever comes by—including maybe us. Visions of having one's neck quickly broken by those vice-grip jaws can be unsettling. We've ended up becoming strange bedfellows with those who shoot Wolves on sight and wouldn't slow down if one crossed the road in front of them.

Here's where our state and federal wildlife agencies come into play. They are entrusted with the welfare of our native plants and animals, and were it not for them, Eagle, Moose, Pine Marten, and several other iconic Northwoods residents wouldn't be here today. Unfortunately, when fast action is needed, these agencies tend to respond with typical bureaucratic speed. At times their sluggishness supports sound management, such as when they make sure to gather and assess solid data before changing policy. The upshot is that for Cougar to be recognized as a resident species and gain protection and other management benefits under the Endangered Species Act—a move I endorse—there needs to be supportive data. The call of a Cougar in the night won't do unless it is witnessed, recorded, and voice analyzed. A photograph does no better unless it's part of a photographic series to substantiate location and time. Tracks and track molds are near the bottom of the list, as tracks are easy to fake and molds can't be verified for time and location. As I see it, the main problem with tracks is that few agency personnel have tracking expertise, so they couldn't positively ID a track if they did see

_____
*For more on Cougar safety, see appendix 5.

one. One wildlife biologist straight-out told me that Cougar leaves no track.

Scat containing hair from self-grooming is considered solid evidence, because the hair can be identified by DNA analysis. It is the most reliable and accepted proof short of a carcass or live-trapped animal. Even without hair content, scat is usually the easiest source of DNA to get. Still, not all Cougar scat yields Cougar DNA, as so few cells slough off of the intestinal wall.

I favor DNA testing because I think it will eventually put to rest the belief of many wildlife professionals that all Northwoods Cougars are released pets originating in Central and South America. This wouldn't be a problem for the Cougars themselves except for the fact that, being a different subspecies than our native animals, they don't qualify for protection under the Endangered Species Act.

Another argument against us having a native breeding population is that it is typically males who disperse and settle in new areas. When these areas have no resident female population, there is, of course, no reproduction. If upper Wisconsin, Michigan, and Minnesota have a native breeding population—and I think we do—it's likely because private groups illegally trapped females out west and released them here. Confidential sources have hinted to me that this may have occurred.

Let's not forget the political situation. Even though an agency might profess objectivity, and even though most of its staff is made up of dedicated professionals, it is politically governed and funded. This plays into decisions involving Cougar management, which is a sensitive issue with the voting public.

At the moment, such concerns are the last thing on my mind. Lety, clearly as excited as I am, asks if I want to see where the Cat crossed the road, and we head down in that direction.

There could be one or more Cougars in the area, as there has been a cluster of reported sightings only thirty miles from here. Three neighboring states west of us have just had their first confirmed sightings, so it is only a matter of time for Wisconsin. Those Cats are probably dispersing eastward from Cougar population centers in Colorado and the Black Hills. This makes a third possible source for our resident Cats, along with abandoned pets and those illegally trapped elsewhere and released here.

To be sure we catch the track, Lety and I start looking a generous quarter mile before the place where the women stated they saw the animal. Each of us takes a side of the road, and we work our way slowly up to the boat landing. This section of the road is hard-packed sand and stone; only its very edges are soft enough to hold a clear track. In order to maintain perspective so we can catch sign of movement rather than only scouring the edge for an actual track, we stay a pace or two off the road.

"We're looking for a needle in a haystack," I comment after we've covered about half of the distance. "We're coming up with a lot of sign of other animals, and still we're probably missing more than we're catching."

Taking a look around, I realize this area doesn't have the look or feel of a Cougar corridor. There is not proper cover, and if the Cougar did cross around here, she'd run smack-dab into the lake. "I wonder," I say to Lety, "if we ought to be searching farther down the road, where there is more favorable Cougar habitat."

"And who knows what a 'quarter mile' means," adds Lety. "She could've crossed farther down."

As we're talking, it starts to drizzle. We're concerned about rain affecting the tracks and we don't know how much territory we have yet to cover, so we decide to go back to the trailhead and grab our bikes. Pedaling slowly, we listen for the song of Cougar's track.

After about a mile, we come to a dense band of Hazelnut and young Balsam Fir that serves as a transitional zone between upland hardwoods and Spruce-Cedar swamp. "Now this says 'Cougar,'" I say to Lety. "It feels like I belong when I envision being Cougar and winding my way through this toward the road."

A dream I had last night just now comes to mind. In it, Cougar, clear as day, walked toward me. I could feel her sense of purpose as she advanced with determination. At the same time, she wasn't forceful or threatening. The way she was centered within herself and moved as a shadow mesmerized me. She came closer and closer, looking around and through me, but not directly at me, so I had no fear.

It is wild that the dream came out of nowhere the night before this Cougar adventure, and that I didn't recall it until this moment. Rather than questioning it or trying to read something into it, I'll just accept it and stay in the moment.

Just as I finish my sentence, I spot directly ahead of us the tracks of a vehicle that pulled slowly off to the side of the road. "Notice how they veered to the right," I remark, "as though they were looking at something off to the left."

The tire tracks match the description of the vehicle the two women were driving. I stand where the vehicle stopped, become the women, and look off to the left. There's the Cat just off the road, looking back at us. I then become the Cat before the vehicle appeared. Approaching the road, I stop beside the big Red Pine to survey the roadway before stepping out. I hear a vehicle coming and wait until it appears from around the curve. It is quiet and approaching very slowly, so there is no cause for alarm. A few steps and I am on the road. Looking first to the left to reassure myself that the vehicle is no threat, I angle off to the right to skirt the open wetland on the other side of the road. Across the road and safely into cover, I pause to look back and check on the stopped vehicle.

Returning to my own identity, I go over to where I, when I was Cougar, came down the bank and onto the road, and there I see her track.

My dream, the Cougar mystique, the controversy around her return, my first-ever Cougar track—all at once it hits me like a shock wave and leaves me with a heady, almost giddy feeling, like the afterglow of an adrenaline rush.

"Tamarack," calls Lety from a few bike lengths back. Her voice breaks through and brings me back to the moment. She is on the other side of the road looking down at something. "What does this look like to you?" she asks.

I work to contain my excitement so that I can be fully present with her and what she has found. As her mate, I realize that her discovery is mine, just as mine is hers. We've gotten accustomed to the synchronicities in our lives, so I'm hardly surprised that we both came upon something at the same time.

The track she found is deep and clear, even though it is mostly filled with fluffed-up sandy backwash from the Cat giving her paw a flip when she lifted it. No words are spoken as we share a moment of gratitude.

We go to look at my track, taking a wide berth so as not to disturb the possible trail between the two tracks. Mine is hard to see—it is just a rearrangement of pine needles on hard-pack. I trace it with my finger.

Stepping back to gain perspective, we relax and let the tracks between hers and mine materialize out of the random needles and sprinkling of pea-size gravel. The trail is so vague that it appears and disappears, and with a fresh look it appears again. But there it is, with each print methodically placed. The two tracks we initially found turn out to be the first and last of her road crossing.

We mark the tracks with sticks pointing to them from the road shoulder and head for home.

"How sure are you that the tracks are from a Cougar?" asks Lety.

"About 79.3 percent," I reply. She knows I always hedge my bets, so it doesn't close me off to other possibilities. If other sign or other observers confirm what we think we've seen, my degree of certainty will rise, I tell her. Of course, I add, the reverse could happen also.

A little further down the road we meet Rick, my assistant, who is on his way out to the Lake. "What've you been up to?" he asks.

"Oh, we've just been noticing some fresh trails crossing the road," I respond matter-of-factly while giving Lety a discreet wink. "There were several Deer, a few Fishers, a couple of Bears, a Cougar, Coyotes. . . ."

Rick responds with a deadpan look, just in case I'm pulling his leg. Lety and I can't keep straight faces anymore, so we tell the story.

"Can I see the tracks?" he asks.

He has trouble seeing mine, and I find out that I do too if I stare at it. When I move around to change perspective and look afresh, it reappears.

"How about asking Freddy to come and check it out?" I suggest to Rick.

"I was thinking the same thing," he replies.

Rick's mentor, Freddy Kenbridge, learned tracking as a kid by having to hunt to help feed his family. He went on to become a forensics and wildlife tracker, and after he retired, he became a tracking instructor. He is one of the few people in the country who can make out and age the tracks of a vehicle that has pulled out onto an asphalt road. For years hunters have called on him to track down wounded Deer that they've given up trying to find. I look forward to hearing what he has to say about our tracks.

Unfortunately, I wasn't able to join Rick and Freddy at the site. "He thought it was a female," says Rick, describing Freddy's investigation. "He said there's also a small chance it was an adolescent male around ninety pounds. He helped me see where she came down through the Firs and sat beside that big Tree, then stepped out on the road. He didn't have any trouble seeing the tracks between yours and Lety's, even though they were on the hard-pack in the middle of the road. I could kind of see one if he pointed it out, but I wasn't sure. Then he showed me where she turned to look back at the vehicle after she got across the road. I couldn't see anything beyond that, so he showed me what to look for and I could follow her trail into the woods on my own."

"He said Cougars move through the vegetation like a Fish through water," Rick adds, "and that's why they can be so hard to track. After they pass through, the plants close in behind them, like they were never there."

While Rick tells his story, I realize the kinship I feel for Freddy. I don't know many people who've so attuned themselves to the song of the track. Freddy said this was only the second Cougar he has tracked, yet he knew. Here was intuitive tracking at work. Awesome tracks with toe spreads of nearly five inches were only a few pages of the story he read there. No words that I could muster would adequately convey the images of Cougar that probably helped Freddy at the site: impressions painted in his ancestral memories by his tracking Ancestors. No explanation of mine could get across how his once becoming a Housecat and crossing the living room floor prepared him to see a Cougar cross a road the day after the fact. Like my dream, I can only accept this event. At the same time, I am awed by the realization that everything I know and have yet to learn about tracking is already a part of the shared human experience.

Six days later, Kip and Justin, both former students who have been on staff for a couple of years now, tell Rick and me about some odd-looking scat they came across along the side of a dirt road about a mile from the Cougar crossing.

"Let's go check it out," I suggest.

"Didn't you guys say it was right around here, just west of the snowmobile crossing?" Rick asks them when we come up empty-handed after scouring the roadside. We go down another hundred paces, and still nothing.

I flash back on an afternoon a number of days ago when the "quarter mile down the road" Cougar crossing turned out to be a mile.

Looking over the area to see where I might be traveling if I were Cougar, I'm drawn to the brushy edge on the other side of the narrow bog that the road cuts across up ahead. I run over there, find the scat, and wave my arms to signal the others to come join me.

"What do you think?" I ask as they gather around.

"It definitely has the tootsie roll look," says Justin. He is referring to the characteristic grooved rings on feline scat that segment it and make it look like it gets repeatedly squeezed on the way out. This makes it easy to tell at a glance from canine scat, which usually has a smooth surface. The scat we're looking at is about two-thirds the diameter of an adult male Cougar's. The size fits with Freddy's appraisal of the tracks he looked at as being made by a female or immature male. The plug, which is the first portion of scat to emerge, is in the typical cone shape and it is dark, dry, and compressed, just like the segment that followed it. This indicates that it sat in the colon for a while and got dehydrated. The rest of the scat is soft and a light, smoky brown color, which says it is from a recent meal. The inconsistency is not typical for a Cat. Being pure carnivore, they have digestive systems designed to pass food quickly through the body, which results in a fairly uniform scat.

Some Cats deposit their scat in small depressions called scrapes, which they scratch out for the occasion. Other Cats are meticulous about covering their scat, which may be to disguise their presence. Still others leave their scat prominently exposed, which to me looks like either territorial markers or dominance indicators.

A canine makes a different kind of scrape, by kicking with his hind feet after he has deposited his scat. This is done primarily by males to mark the spot with scent from glands between his toes. Occasionally he'll kick his scat around in the process, but usually the scrape is beside the scat. Using scat as territorial markers, canines often deposit it repeatedly in the same place—which makes it a great place to study scat aging. Felines rely more on urine to mark their territories, probably because their nutrient-rich scat is a popular menu item for some scavengers. Those of you who've had Cats and Dogs at the same time may have noticed how a Dog loves to go to the litter box for dessert.

It turns out that right along with Dogs, Cougar researchers relish kitty tootsie rolls. I e-mail photos of the scat we found to our regional Cougar expert. His reply is instantaneous—there is a good probability that what we found is Cougar, and could we please send him a section right away. "It's a little stale," I reply. "A fresh piece might be more flavorful."

A week passes and Lety has just returned from her Ojibwe language class at a nearby reservation. She tells me that they are learning the Ojibwe names of the local animals. "Most of the students are Ojibwe and they just memorize names," Lety says, "because they don't know most of the animals. Not even the common Birds."

Her words lay heavy on my heart. The animals' names are just the tip of the iceberg. Since the missionaries and boarding schools started teaching them "a better way of life," so much has been lost—and all in just a couple of generations.

In the evening, Karl Vanderzen, an old non-Indian friend who lives with the local Ojibwe, stops by. He brings up the topic of *Mishibizhiw*, Ojibwe for Cougar, because he has heard talk of sightings in the area. He says that Mishibizhiw, who was once an important relation to them, is now all but forgotten. "There's an old Ojibwe prophecy," Karl relates, "that tells of the time when she'll return from the north."

I ask where he heard the prophecy.

"I'm as excited about Cougar coming back as you are," he says, "so I went to an Elder to ask him about it, and he told me about the prophecy. It says that Mishibizhiw's return will be a sign that a new group of people are going to step forward and become Earth Mother's guardians."

Cougar's shadow lingered just like Wolf's did after he was exterminated. The prophecy, I realize, is very similar to the one that links the fate and fortune of Wolf and human. All of a sudden I'm hit with the same feeling I had when Lety and I first found the tracks. What is going on? Again the feeling of synchronicity with the dream, the two women, the tracks, the scat, and now the prophecy. It resonates with my life's calling to work with others in remembering our ancient ways of living. However, it is just a beginning, only a portal. Like the sun suddenly breaking through the clouds and flooding a meadow with light, I'm coming to realize my kinship—the human kinship—with Wolf and Cougar. The three of us are apex predators who have lost our place in the Hoop of Life as guardians of the Relations, and now together we are returning to resume our role. And in my lifetime!

Now when I come near a ledge or pass under a large overhanging branch, I look up to see if Cougar might be there. When I follow a Deer trail, I realize that I may be stepping in Cougar's footprints. My relationship with this wilderness, and with myself, has forever changed.

Were it not for a single track, it may not have happened. I am grateful for the budding relationship I am privileged to have with our resident Cougar. My sense is that she is a female. She moves with the confidence and savvy of a mature, native animal, and she is probably too small to be a mature male. It is unlikely that an immature male would have dispersed this far from his likely birthplace in the Black Hills or the Rockies. Cougars are natural loners, so she probably doesn't mind having the wilderness to herself. That could change next spring if she has cubs, and I would consider it a great honor to meet them. But that is not my focus; what really matters to me is that the forest shadows again embrace Cougar.

Two weeks have passed and Mark, our staff carpenter, is on the way out to camp to help build a wigwam. He cuts off the trail to relieve himself in a brushy strip of edge between the Pine and Maple forests and comes across a dead yearling Deer lying beside the trunk of a fallen Tree. "It's a strange-looking kill," he says to Rick when he gets to camp. "There's no blood, and the animal's partly covered with debris."

Along with several staff, there are seven students working on the lodge, and they all want to go and see the kill site.

"Let's wait until I get back," suggests Rick. "I'm going to check in with Tamarack about the site, and I'll grab my camera so we can get pictures before it is disturbed."

Lety and I, already on our way out to camp, meet Rick on the road much like we did a few weeks ago when we found the tracks, only this time he has the story. And wouldn't you know, he deadpans it.

While he is filling us in, I feel a smallness creep over me. Not the smallness of insignificance, but of realizing that I am not the master of my life. It is as though I am living in the shadow of something: of an old, old story, of a long-ago people, of a mysterious animal. I get the sense that I am moving within some greater movement that gives me identity and purpose.

While Rick goes for his camera, Lety and I stalk in to the site. We want to pick up on any sign that might be disturbed later by the group, and we want to figure out how to guide them in to the site with the least impact. We approach from the north, to best catch the shadows of ground disturbances. There is no better time of year to pick them up than now, when the sun sits low in the southern sky.

As is our custom, we completely circle the site in a sun-wise direction, and we stay about twenty paces out from the kill. This gives us a good idea as to who came and went from the site, and when. Using the fallen Tree as a pathway to the kill, we're able to get up close without disturbing any ground sign.

"It looks like the kill site may not be the place for us right now," I say to the group at camp. "The kill looks to be Cougar's typical neck-choke ambush, and Cougar's return is not only a gift for all the Relations, but it could be a sign of things to come. Cougars usually stay very close to their kills, and it looks as though this one bounded off when Mark got too close for comfort. There could be a reason she is hanging out near us, and perhaps we can best honor that and help her feel welcome by leaving her in peace. Maybe we can all go take a look at the kill site in a few days."

Everybody agrees.

"How did it feel coming onto the kill site?" I ask Mark.

"It was odd," he replies, "like I was excited and scared at the same time. I didn't know if the killer was still hanging around or not."

Many people report feeling the way Mark did around a large Cat's kill site. Wolves hunt in a pack and tear at an animal to bring him down. They sometimes start eating even before he is dead. The chase and struggle leave a lot of obvious sign, like torn up ground, blood, and patches of fur. The story is easy to read, which puts the mind to rest.

Not so with a Cougar kill. Like all other Cats but Lions, Cougar is a solo hunter, who has no one but herself to rely on. Chasing is not a good strategy because there is no one to relieve her, nor is there anyone to intercept an animal zigzagging to escape. Even if Cougars were to hunt cooperatively, they couldn't maintain a chase, because they're not designed for distance running.

Instead, Cougar is a master of the pounce. She lies in wait, and at just the right instant she springs on powerful, oversized back legs and her long, limber body uncoils like a spring. In one or two bounds she is on the animal's back and suffocating him by clamping her jaws around his throat. It is often a very clean kill, with little or no visible blood, not much sign of a struggle, and of course no sign of a long chase.

Imagine chancing upon a dead animal when the cause of death isn't obvious. If the carcass hasn't already been feasted upon, the site could look all the more mysterious. Cap that off with the feeling that somebody is watching you from behind, which could be the case with Cougar, as she tends to hang out close to her kill—often very close. And if she were there, the odds are you wouldn't see her. Even if you did, she might trust so much in her camouflage that you could look directly at her, and even take pictures, and she'd still sit tight.

It is now three days later and Lety is guiding the students and three staff members into the site the way she and I initially approached it. To minimize disturbance, they "Wolf walk." Lety asks them to take the fallen Tree trunk to get up close to the kill. "Keep silent and stalk in," I suggest. "It'll help sensitize you. Maintain a state of broad awareness, so that you can gain perspective and not get caught up in details. Later we'll take a close look."

One by one, they creep up to the kill site. After everyone has had their turn, Lety motions to move away from the site and we sit in a circle to share observations. Nobody speaks above a whisper.

"What jumped out at you?" I ask.

"It's obvious there was a scuffle," says Ingrid, "but I didn't see any mark on the Deer. It looks like he just fell over."

"I saw some matted hair around the throat," adds Ryan. "And his rear end was chewed into a little bit."

"So how was the Deer killed?" asks Matthew.

"I think Ryan answered that," replies Rick. He explains how Cougar suffocates her prey, a fact which we would normally let them discover through their own research. But they need to know it now—if this were, in fact, a Cougar kill, of course—in order to get a feel for the drama played out here.

"Are you getting a feel for why Mark felt edgy when he stumbled into this?" I ask.

Several people nod.

"Did you see the hair on the other side of the log?" Hans asks.

"There wasn't much," responds Kip. "And a couple of those tufts didn't look like Deer hair."

"Yeah," adds Scott, "it was lighter colored and finer. What do you think it was, Tamarack?"

"We're sending a sample to a lab for analysis," is my only reply.

If I gave any more information, it could dampen their inquisitiveness. At worst, they would take my word as gospel and quit questioning altogether. My role is like that of a mystery story writer, who gives clues but doesn't solve the crime until the very last page. The students' roles are those of master trackers, who are no different than master detectives. There is always a twist in the plot, so just like the writer, the detective and tracker have to hold out on forming a conclusion until the very last clue is followed up.

"It looks like whoever killed that Deer didn't leave any tracks," states Kip. "How can that be?"

"Any ideas, group?"

"Well, if it was a Cat," says Monique, "they have padded feet with retractable claws, so she could walk over these leaves and needles without making much disturbance."

Here is a good example of how turning a question back to the group can help them discover how much they already know collectively. This is clan knowledge at work. Each person carries a bit of information, which when pooled together, displays a collective intelligence that could make a believer out of even the staunchest individualist.

"There's something important about this kill," I remark, "that is an important aspect of all kills, and that is 'Why here?' If you want to learn how to be an effective hunter, or just how to find kill sites to study, this is the question to keep asking yourself."

There is a long silence.

"Become the predator," I suggest. "We're not sure a Cougar made this kill—it's still 'Be as a question.' But let's assume for this exercise that it was Cougar. Step inside her so that you can come to know what she considered in choosing to lie in wait at this particular spot."

Lety, Rick, and I give the students some core information to aid their becoming. We explain that Cougar hunts by ambush, so she needs to envision where her prey is going to be and when.

Ambush hunting takes a brilliance that canines, who run down their prey, don't have. This aptitude is one of the reasons Cats thrive as loners. Even the House Cat, as affectionate as she might be, still listens to a voice that is very much her own. It is a voice that often puts her at odds with her owner. Cats are the epitome of self-reliance and independent thinking. They're superbly attuned to their environment, knowing their place and exactly how to function within it.

We humans are unique predators in that we have the brilliance of both Cat and Dog. We can execute a coordinated chase and run down even the Antelope who is able to escape Cheetah. Or just one of us could take the Antelope by determining the time and place to lie in wait for her to come by. Perhaps our dual ability is the reason many trackers feel a kinship with both canines and felines, as well as the reason Dogs and Cats are our most popular pets.

The students struggle with the question of "Why here?" which tells me we need to approach it in smaller pieces. I use the break-down approach regularly in problem solving for both students and myself. After all, a large project is no more than a combination of small projects. In this case, I begin by asking if anyone saw the scat that we nearly stepped on while circling the kill.

Two people nod.

"How about the munched-off Sedge and Raspberry?"

Nobody responds.

"Ah, there's food here for the Deer," says Ingrid in her thick Swedish accent. "They come here to eat, and the Cat knows that."

"Whose scat is it?" I ask. "And how old is it?"

We go to look and they recognize it as Deer scat. They're all over the board with its age, so I ask Rick for his input. Again, they need this information now or else we'd encourage them to figure it out on their own.

"It's probably from this morning," Rick says, "judging by how shiny it is. The pellets are firm and well-shaped, which means they're high in fiber. Deer have had to switch from tender new growth to tough grasses and twigs. The Raspberry clippings look pretty fresh too, like they could be from last night or early this morning. I'm going by the weather and how much they're shriveled and oxidized."

"Yeah, food is one reason they're here," I state, "but there's also food elsewhere. There are a couple of other very important reasons for Deer traffic through this area."

"Well, it's different," Monique tentatively begins. "On one side is a Maple woods, which is very open underneath, and on the other side is Pine, which is pretty open too. But here in between it's thick with brush and small Balsams."

"Why would that attract them here?"

"Because they have shelter?" offers Ingrid.

"A very important consideration for a prey animal," I reply. "This is edge habitat, which is typically rich and brushy. There is still another reason animals are drawn to edge," I continue, "and this edge is a good example. You've been here for half a turn of the seasons now and know the area well. What have you noticed that would draw animals to this strip?"

The students can only dance around this one, which I expected. My intent was to plant a seed to pique their curiosity. Without a seed, the discovery process has no place to begin, and students often flounder and get frustrated. In the coming days, with a few additional clues from us if necessary, they'll come to realize that this edge is a natural corridor for Deer. They travel back and forth from the lakeshore just north of us to the highland on the far side of the big bog to the south. They use this strip so much that someone is bound to discover places where the trail is worn in enough to be obvious.

"There are other tracks and sign here," I continue, "only they're much harder to see than what Deer left us. A big Cat appears to have passed through here heading south and went right by the Deer scat we just looked at, and there's other Cat sign around the kill site."

"Can you show us some?" several ask at once.

Rick and Freddy Kenbridge were here at the site earlier in the day, and again I didn't get to join them. Rick shows the students where Freddy had pointed out two Pine Trees the Cat marked by rearing up and raking them with her claws. The scratch marks don't look very remarkable to anyone, considering that they were made by an animal with razor-sharp claws and the strength to bring down a prime yearling buck. Someone says it looks like a halfhearted attempt.

Maybe it is, and at the same time I suggest that there could be other possibilities. With their curiosity up, the students will explore and hopefully discover that Cats don't like marking conifers, because the pitch coats and irritates their claws. In this case, the only nearby Trees were conifers, so she had no choice but to use them if she was going to mark at all. In addition, the Cat could be an adolescent, who wouldn't have the hormonal drive to leave a prominent mark.

That evening, Rick, Lety, and I look closely at the hair samples we collected for lab identification. "That's Bobcat," is my instant response. The hair has sooty overtones rather than the clear sandy color typical of Cougar. It'll be a while before we find out, as our regional DNA analysis lab is notoriously slow.

"Be as a question," I tell myself, and I go over the story of the kill. Quick notions like mine are based on prior knowledge, which can't be trusted because they're based entirely on past experience and don't take new information into account.

Two things stand out: the Tree scrapes and the Deer. The scrapes could appear halfhearted and be lower on the Trees than a Cougar could reach because they were done by a smaller and weaker animal than a Cougar, such as a Bobcat. However, the odds of a Bobcat killing—or even trying to kill—a Deer in his adolescent prime, without him being either injured or bogged down in snow, are close to nil. Perhaps a shedding Bobcat passed through after the kill was made. Or the hair could have come from an immature Cougar. Cubs typically retain their Bobcat-like markings for about six months, and occasionally up to a year.

Thanks to remembering to be as a question, I now have a range of possibilities to fit all the conspicuous sign. One possibility resulting from the questioning process is that our Cougar isn't a mature female. I wouldn't be at all disappointed, because if she was a youngster she has a mother, which could mean we have a mature breeding female in the area *and* more cubs. Here is a prime example of how questions can pay off better than answers.

Confusing sign seems to be par for the course for Cougar. People commonly mistake her call for that of a chirping Bird. The incessant yowl of a mature female in heat—you have to hear it to believe it—falls somewhere between the hiss of a leaky tire valve and someone trying to scream with a sore throat. It'll leave you scratching your head unless you've been around a Housecat in heat. Not even night sightings based on Cougar's eyeshine are reliable. Typically thought to be bright yellow-green, their eyeshine can vary from blue-green to gold, depending on Cougar's age and the angle, source, and intensity of the light.

A week has passed and I find myself deep in a galaxy of big, cottony puffs of snow. They drift on a gentle easterly breeze and give my cheeks a refreshing ting. What a lyrical way for the Earth Mother to lay down her protective blanket for the coming white season.

At first light the next morning, I step out to see that the wet snow has settled to ankle depth and that the sun promises to rise bright and inviting. Rick and I had planned on going out to camp to do some awareness exercises with the students. Considering the ideal tracking conditions, I ask if he'd like to take advantage and leave right away. Here is one time where it just doesn't pay to be as a question, because I always get the same enthusiastic answer. Some people are so predictable! But I ask anyway, just for the formality.

I have Rick drop me off about three miles from camp, right around where Justin and Kip found the Cougar-looking scat a few weeks ago. Rick goes on to track a family of Deer he is getting to know. I'm standing near the southern end of a bog that runs northeast a mile and a half and connects with a lake about two-thirds of a mile long. Together they create a wetland barrier that significantly affects the east-west travel of animals in the area.

As I make my way through the Tamarack Trees bordering the bog, my wool shirt turns to tawny fur and my boots to padded paws. My cognitive functioning switches from rationally based to sensory-intuitive, and my hunger for skill and knowledge transforms to a focus on sign and a lust for the hunt. Thoughts of the world I come from disappear and I become a creature of the forest and the moment.

At the bog's edge, I pause to listen and sample the air. Hunger is my driving force—I need to get to my hunting territory across the bog. However, I'm uneasy about crossing such a long, open expanse.

On the far side of the bog, I spot a wooded peninsula jutting out, and from this side another peninsula reaches toward it. Together they cut the distance across the bog in half. What a break! I envision the lay of the peninsulas and see a trail running the span between them. My heart quickens and my senses keen as I head due north to intercept the trail.

When I reach it, I'm drawn instantly into communion with the now. It isn't the picture-perfect pathway worn deep into the velvet Moss and shaded by golden Labrador Tea that did it, but rather the paw prints of a very large Cat. I reverently slip my feet into the tracks, and as one we tread quietly to the far peninsula. We think and feel the same; our only separation is the brief time between predawn when she came through and now when I become her shadow. In the scheme of things, the time difference is insignificant.

Every step is deliberate, as it brings the opportunity—and need—to gain new perspective on my surroundings. Along with threats, there is always the potential for an unexpected meal. Then there is the trail itself, which I am memorizing for the next time I pass this way. It is not just a way to get across the bog that I'm retaining. I'll be able to replay my crossing like a virtual reality experience, complete with sound, smell, and feeling.

To maintain my human tracker perspective, I slip in and out of being the Cat. Sometimes I literally step in her tracks, and at other times I'm paralleling them, either to gain perspective or to preserve the trail.

Being completely exposed, along with feet soaked in icy slush, is not this Cat's favorite way to go, so we move with extra caution. We slow down even more when we reach the highland, which is blanketed by a solid canopy of Hemlock and Cedar and choked underneath with young Firs. This is a great corridor for animals moving north and south, and we're curious to know who has come by recently. We see sign of Fox, Coyote, Deer, and Snowshoe Hare. Satisfied, we squat to take a pee and head inland.

I linger a moment to smell the pee, which is the same light color and very mild urine essence as that by the Deer kill site. Definitely not an adult male. Scratch that—he could be a neutered, released pet. Anyone who's caught a whiff of an endowed tomcat's litter box knows pretty well what a male Wildcat's urine smells like. When you come across that scent in the woods, follow your nose and you'll likely find a marking post.

The snow is melting and the students will be expecting me soon, so it is time to travel fast. My paws revert to hands and feet and my fur transforms to green wool. In order to cover a lot of territory quickly and keep track of the trail, I run a zigzag pattern. When I want to find the trail and head the wrong way, I feel that I'm getting colder, and when I turn toward the trail, I feel that I'm getting hotter. My overall direction is guided by the tell-tale maneuverings of a Raven up ahead, who tells me that I may be catching up with the trail-maker. And then there are the subtle voices constantly at play that don't register consciously. Still, they influence every thought, feeling, and movement.

It might appear to an observer that she and I are separated by half a mile and a species barrier, and that we are each responding to different input. However, it is clear to me that we're both moving in the same direction, to fulfill the same primal needs for knowledge, comfort, and nourishment. We are related; our movements are one movement.

In an open, snow-covered grassy area under lofty Pines, I come upon a scrape of Cougar's that is as long as I am tall and almost as wide. Wanting to gain perspective on it, and at the same time not wanting to disturb it, I stay back and listen to its voice. There is no scat. Maybe she squatted to pee and marked the spot much more boldly than the one at bog's edge because here she is not as distracted. No, distracted isn't the right word—she was the opposite of that. She was taken up by the boisterous chorus of all the Relations who used the corridor. Here it is open, level forest with no cover and little sign of animal activity.

So why such a big scrape for only a urination—especially such a weak-smelling one? Maybe this is an immature male who doesn't have the hormonal charge to produce strong-smelling pee. He might squat because he doesn't have an established territory to mark, so he has no reason to spray a scent marker. Still, he feels frisky enough to do an aggressive scrape. Or is this a female doing something atypical of her gender?

Regardless of age or sex, the song of this Cat's track keeps repeating the same verse through these Pines: *she moves with calm, she moves with confidence.* Her long, determined stride shows no confusion or hesitation, and certainly no fear. She is alert, for sure, but not on guard or particularly inquisitive. She may be hungry, but she doesn't look pressed to find something ASAP.

Her two-boot-length stride is impressive—obviously a big Cat! I haven't seen a Bobcat stride approach that—it is usually around one boot length or a little more. I wear a size nine PAC boot, which is about twelve inches long. And this Cat's straddle is a consistent hand-length wide—around seven inches—which again shows a purposeful and self-assured animal. If this were a Bobcat, he'd easily have the broadest chest of any Bobcat I've come across.

I generally don't carry or use a measuring device, because it gives a clinical edge to my relationship with the natural realm. I couldn't imagine trying to deepen my relationships with my woman friends by taking their measurements. It is the same for me with my nonhuman Relations. There is a rational component to the natural realm; however, it is not rationally based. Calculating and charting pull me back into the old rational mindset they ground into me during my college days as a Wildlife Management major. Still I recognize that there is a place for the studied approach; only when we use it, let's keep in mind that it tends to create distance.

The common native way of measuring is to use a part of your body, such as your finger, arm, or foot. It personalizes the relationship with our plant and animal kin. When I compare the width of my hand to the width of someone's paw, I'm meeting them on common ground.

This morning's packable snow, along with the sparse vegetation here under the Pines, is ideal for track registration. The front pawprints of this Cat show the round shape, with inner toes projecting more forward than the outers, which is typical for Cats. The pullout, which is the scrape the front paw makes when lifted up through the snow, is wide and round—typical also for Cats. I compare the foot pad to my fingers; it is three wide, which is about two and a half inches. She does not direct register, which means her back paw doesn't come down precisely in the print of the front paw. This is a characteristic of a long-bodied animal like Cougar. Bobcat and Lynx usually direct register in snow. The back paw of the Cat I'm tracking falls just behind the front and there is usually a slight overlap. Her rear paw tracks show she is slightly duck footed—another Cougar characteristic.

"Could this be Lynx?" I ask myself. For the deep snow of the Far North, they have built-in snowshoes, also known as oversized paws, which fall in the size range of a Cougar's. However, a Lynx averages only one-third the weight of a female Cougar and has a shorter stride. Cougar's toe pad prints show clearly in snow, while I've been told that

Lynx's are obscured by fur. In these tracks, the pads are clearly defined. Yet I bet it could still be Lynx. In these conditions, the wet, matted fur between Lynx's pads might allow the pads to register. Lynx's trail, like Bobcat's, meanders, where this one is fairly straight, which is typical of Cougar. On top of that, Lynx is very rare in this neck of the woods. In fact, I've yet to see sign of one.

Here the Cat and I part. She continues on to the northeast through new-growth Aspen that cover the rugged glacial landscape, and I head due north to join Rick and the students at our lakeside camp. At the same time, I stay attuned to the song of her track and continue on with her.

I won't mention anything to the students about my morning. It is a deeply personal experience and I want time to be with it. At the same time, I know I'm not intended to keep the story to myself forever. It is through story that we learn, which is why I believe the story of the hunt can be an even greater way for a tracker to serve his people than the fruit of the hunt. One day, I will share the story of Cougar with the students, so that they'll be able to slip on tawny coats and glide on padded feet.

Acknowledgments

When tracking, I am an observer. The animal I trail calls the shots: she taunts and mystifies me as I strive to reach her. She is the clever one—the one I must humble myself before and know intimately if I am to have any hope of outwitting her. Without me, her life goes on as usual. Without her, I am a tracker without a trail to follow. As my teacher, she has gained my respect and admiration with dazzling displays of deep intelligence. As the source and inspiration for so much of what I know about tracking, she has my deepest gratitude.

Like you, I am a born tracker. Still, I needed to learn the language of the track. Training from indigenous peoples, particularly the Ojibwe of my area, has been invaluable, as has the support of Australian Gandangara Aboriginal Elders. May I never forget their gracious ways.

Even though I know my awareness is not limited to what I alone observe, I am still amazed at how my world expands when I work closely with student trackers. The more I am able to see through their eyes, the more perspective I gain and the better tracker I become. Their adventures make it possible for me to revisit lessons forgotten and missed learning opportunities. The crucial role these apprentices play in my awakening makes my story theirs as well.

Writing is said to be a solo occupation, yet I have found that it takes others to provide support and ensure a quiet environment. I am blessed to have both granted by a gracious staff, who have become my extended family: Susan Bean, Evan Cestari, Bärbel Ehrig, Marcus Gardner, Stephanie and Justin Lowthorp, Stephanie and Sam Ross, Carl Rounds, Rachel Sems, and April Winchell, along with past staff members and volunteers. Special recognition for research and field study goes to my longtime assistant Chris Bean, and to staff members Matt Nelson, Tim Nelson, and Thomas Seibold. I am ever grateful to the donors who support my writing and research through the Old Way Foundation.

Even though a story might seem to flow effortlessly, storytelling is a demanding craft. While the story itself is important, the storyteller's ability to be present and engaging is far more essential. My many-gifted chief editor Jessica Leah Moss worked with me through the entire project and never shied away from coaxing the best out of me. Editor Margaret Traylor, a gifted storyteller herself, did much to help this book capture the allure of a live storytelling. My gratitude to interns Laura Ofstad, Rachel MacFarland, and Matthew Neall, along with readers Tim Cronin, Stacey Ginsburg, Natalie Morse-Noland, Claire Shefchik, Robin Sneed, Luke Somers, and others too numerous to recognize here, for their contributions. Their perspectives helped us find blind spots and smooth rough edges. The book's final form, with its balance of narrative and technical, is due to the visionary gift of my agent, Rita Rosenkranz. From title to cover design and layout, the inviting presentation is the credit of the creative team at Inner Traditions–Bear & Company. Managing Editor Jeanie Levitan, Special Projects Editor Erica Robinson, and Editor Mindy Branstetter adopted this project as their own and helped bring my dreams for this book to life. And then there is you, and me, who owe a gracious nod to every person recognized here, as it is all of them together who gave you a well-crafted book, along with making me look pretty good as a writer.

The gestures and inflections that help paint a scene are lost when a story is transcribed. I saw a way to bring some of that back to the stories in this book when I came across an illustrated note card. Unfortunately, the only contact information on the card was *Webster*. It took more doggedness than tracking ability to find the artist, who to my delight was both interested in illustrating *Entering the Mind of the Tracker* and a pleasure to work with. Thanks to Mark Webster and his classic pen-and-ink field sketches, this story possesses some of the life of a live storytelling.

One person stands out as being intimately involved in all aspects of creating this book. My mate and regular tracking companion, Lety Seibel, transcribed most of the text from live recordings and lent her talents wherever needed. Her passionate presence appears on every page.

APPENDIX 1

For Educators, Naturalists, and Tracking Instructors

This book is a call to a new way of learning, where natural rhythms set the pace rather than deadlines and semesters. Where nature is the classroom, and example and experience are the methods. Where the student is motivated by curiosity and passion rather than reward and punishment. Where the teacher serves the student rather than the student trying to please the teacher.

The only thing that makes nature-based and student-centered learning new is that most contemporary people are not familiar with it. Our forebears have practiced it for nearly all of human existence and discarded it only on the doorstep of the sedentary era. When they abandoned their nomadic hunter-gatherer lifestyle and became farmers, herders, and villagers, they began to isolate themselves from nature with fences and regimens. Forgetting that our nature *is* nature, they left behind a part of what it is to be human.

The cornerstone of nature-based learning is team guiding, with the team being the student herself, rocks, plants, and animals, you the facilitator, and whatever situations or discoveries you all find yourselves involved in. All work together synergistically, with you playing an auxiliary role as catalyst, guidepost, and information source.

That said, your most vital task is actually to *refrain* from doing something you might be practicing as gospel: answering questions. Receiving an answer, the student feels satisfied. She loses motivation to explore further and goes on to something else. She may end up with an impressive accumulation of knowledge; however, it often lacks depth and she ends up shy on insight and perspective.

To quote from one of this book's accounts, "A good question is worth a handful of answers." Instead of answers, the educator deftly replies with questions designed to guide the student on her quest for her own answers. The process fuels her passion: it enlivens her senses, stimulates her mind, and enriches the experience with emotional spice. Once they get used to the approach, both children and adults usually respond enthusiastically, as it is their natural way of learning.

It may seem as though asking questions would be easier than coming up with answers. I usually find the reverse to be true: guiding a student draws upon my creative resources more than teaching. I could give pat answers all day without skipping a beat, but to truly listen to someone and craft a question that steers her right back to her own path of discovery is a profound responsibility that can be tremendously challenging.

At the same time, I find it tremendously rewarding—much more so than my experience with standard teaching practices. What a joy to watch someone come to realize that there are no real answers—that one question leads to another, and then another. The root of the word *question* is *quest*. The goal is no longer mastery, but to pursue a continual journey of discovery—an endless tracking adventure. My ultimate goal as a guide is achieved when the student of the track becomes a student of life. She then no longer sees herself as my student, but has firmly placed herself in the hands of the teachers all around her.

Complementing
the Naturalist Approach

Observing and studying the ways of animals and plants can be a beautiful and rewarding experience. It has enriched the lives of many by deepening their understanding of the natural realm. And still, this method may not be enough for those who truly want to know the animal they are researching. The method maintains distance, which impedes developing a relationship with the animal. Imagine telling your new lover you would like to grow closer to him by examining him with instruments and doing research on him in books and keeping a distance in order to observe his behaviors. Now imagine how far the approach will get you.

To go beyond following the trail and start tracking intuitively—to start feeling it in your bones—takes crossing the border between yourself and the one you track. You start to think and feel and move as she does. Her nature becomes yours, which is nature itself. No longer does she live in one world and you live in another. No longer do you feel like a visitor.

I have never found a word in a hunter-gatherer language for *wild*. The very concept is foreign to them—it is impossible for them to talk about a natural area or wilderness, as we do. They can't go out in nature. For us, nature is the unspoiled realm that lies out there beyond the borders of our towns and farms; however, they have no towns and farms to go beyond. Nature as we know it is their entire world.

And it is the same for the intuitive tracker. *Entering the Mind of the Tracker* will guide your students to kindling an intimate relationship with the natural world, much like the one they have with their closest friend. They will come to know the ways and secrets of the plants and animals as though they were their own. The lessons, the awareness, and the examples will help transition them from observer to immersion, from studying the animal to becoming the animal.

Ways to Use This Book

The world of the tracker will be new to many of your students. To familiarize them with intuitive tracking concepts and terminology, along with aboriginal ways of perception, consider going over the glossary terms with them before starting any work in the book.

The accounts in the book introduce and develop three skills that are core to tracking and in-depth environmental studies:

- **Envisioning:** Re-creating the scene in your imagination to replay the events that led up to the kill or other disturbance the student discovered. This is personalized living history—bringing the past alive and into the present—and it is a primary tool for understanding cause and effect relationships.
- **Becoming:** Stepping into the shoes of the animal in order to know his feelings and motivations. This engages the whole person in the discovery process by involving physical, mental, emotional, and sensory faculties. It works as a key to transforming learning experiences from intellectual to experiential.
- **Questioning:** Opening to all possibilities, whether or not they make sense. This approach reveals the beauty and mystery in what might otherwise pass as commonplace. It frees the student from the limitations of current knowledge and ability. He becomes his own guide and motivation, which personalizes the learning experience.

These skills are introduced early in the book and developed gradually as you progress. Find the chapters that offer training in a particular skill by consulting the synopses below.

A feature in all but a couple of the chapters is the modeling of a traditional mentoring relationship. Several noteworthy examples are identified in the synopses. Mentoring techniques are laid out in detail and presented in actual tracking scenarios. This gives a true-to-life feel for how the techniques work in the field, along with making it easy for you to adapt them to similar scenarios with your students. You might want to point out that the characters are novices just like them, to encourage them to identify with the characters and progress along.

Those of you with education experience and established programs most likely already have the wherewithal to utilize the narratives and incorporate the lessons. If you are new to the vocation, here are some suggestions to get you started:

- Use the book as the basis of a tracking course.
- Choose particular chapters to supplement field experience.
- Set up exercises patterned after those in the book.
- Personalize the experience by sharing your own tracking stories and asking your students to recount theirs.

Remember that this is not just a tracking book—it is also intended to foster the development of deeper relationships than what nature studies alone can effect. Take advantage of the vignettes of animal lore and natural history by using them for inspiration and as launching pads for exploration. Personally insightful passages can serve as guides for exploring the nature within.

To help you choose the chapter or excerpt that would best serve you and the student, here is each chapter's central theme, along with the lessons embedded in the chapter.

One: Sweet Fern Rendezvous

A training exercise in abandoning your persona in order to give full attention to the trail and sharpen perception. Demonstrates how to reconstruct a scene realistically by combining sign reading, natural history, and the song of the track.

Two: Romancing the Frog

An introduction to envisioning, the chapter shows how different a scene can look when viewed through the eyes of the animal. Demonstrates how the senses can compensate for each other when reading sign. In this case, hearing gives the picture of an event that cannot be seen.

Three: Bear Stump

Thorough instruction in three core tracking skills: envisioning, becoming, and being as a question. Use the story as the basis of a skills workshop. Extensive demonstration of student-centered education techniques, along with examples of how they empower the student.

Four: Dogs Will Be Dogs

Building on "Romancing the Frog," this exercise shows how one sense compensates when another is compromised. Demonstrates how a little bit of knowledge combined with a lot of rationalizing can lead to confusion, and how a guide can help the student find clarity.

Five: Stalking Turtle

Introduces and applies kill site tracking skills to major events such as environmental disasters. Shows how to read their often complex and confusing sign. Includes a tutorial on the essential stalking skill of becoming invisible in plain sight.

Six: How to Learn Tracking from One of the Greatest Predators

Stalking proficiency comes from daily training, and this chapter introduces one of the best and most accessible coaches. Includes a web-tracking tutorial, a study in becoming an animal, and in-depth invisibility training.

Seven: A Winter Riddle

Instruction in scat analysis, along with lessons in how to avoid making assumptions, over-rationalizing, and jumping to conclusions.

Eight: Eyes That Shine

Teaches the downside of acting entirely on prior knowledge, using the example of a time when camouflage becomes a liability. Mid-level training in becoming an animal, along with techniques educators can use to encourage self-motivation and initiative.

Nine: Grandfather Tip-Up

Lessons in the essential tracking skill of perceiving time in the same way as your quarry, and in how to keep personal and cultural perspective from clouding over the track's story. A guide to transitioning from observer-analyst to being in relationship.

Ten: Becoming Wolf

Advanced training in envisioning the scene and becoming the animal. Practical examples of how fast a tracker can—and often needs to—follow a trail. Includes two demonstrations on how to keep mental chatter from muting the song of the track.

Eleven: Old Songs Never Die

A lesson in using tracking skills to reconstruct historical events. Includes evaluations of technical tracking and reading sign to gauge forest health.

Twelve: A Sound Lesson in Tracking

A briefing on sound dynamics, along with demonstrating aural tracking methods. Presents a guided envisioning exercise that can be easily replicated. A clear example of guiding versus teaching.

Thirteen: Eagle Spirit and the Tin Can

Two lessons in backtracking—a challenging task even for accomplished trackers. Covers procedures for backtracking animate and inanimate objects. Includes an exercise in the specialized skill of feather reading.

Fourteen: Following the Flight Trail

All the training from previous chapters brought together in the book's most challenging tracking case, which has very few visible clues. Features a concise course in shadow tracking.

Fifteen: Human Tracking, Bear Style

Training in the finer points of shadow tracking and becoming, based on a human tracking experience. Presents some of the unique characteristics of human movement patterns and sign, along with showing how to learn human tracking from Bears.

Sixteen: The Messenger

A briefing on the tracker's relationship to the world at large, including conflicting personal and societal values, the challenges and opportunities of working with bureaucracies and native traditions. Includes exercises on Cougar movement patterns, hunting methods, and kill site reading.

APPENDIX 2

Demystifying Canine Track Identification

After weasels, I've found members of the canine family to be the most challenging of any mammals to distinguish by their tracks. Novices often find this hard to believe, as Foxes, Coyotes, and Wolves seem to differ markedly in size. At the same time, Justin is not alone with situations like his in chapter four, "Dogs Will Be Dogs," where he says, "Well . . . I think Coyote . . . it looks too big for a Red Fox . . . it can't be a Wolf, unless it's a small one . . . a Dog! No, it couldn't be."

To get a picture of the problem, take a look at the overlap in weight and front track length shown in the table on the next page.* Track length measurements do not include claws, as they often do not register.

---

*All tables and illustrations in the appendices were created by the author from personal and staff specimen collections, field notes, research, and photos.

CANINE SPECIES SIZE COMPARISON

SPECIES	WEIGHT	TRACK LENGTH	IDENTIFICATION TIPS
Kit-Swift Fox (*Vulpes macrotis-velox*)	3–7 lbs (1.4–3.2 kg)	1-1¾ in. (2.5-4.4 cm)	Long legs and body can make him appear as large as Gray Fox.
Gray Fox (*Urocyon cinereoargenteus*)	7–15 lbs (3.2–6.8 kg)	1¼–2 in. (3.2–5.1 cm)	Shorter-limbed and stockier than Red Fox. Looks smaller even when the same weight.
Red Fox (*Vulpes vulpes*)	6–22 lbs (2.7–10 kg)	1½–2⅞ in. (3.8–7.3 cm)	Lankiness makes him appear Coyote size. Leg bones 30 percent less dense than Dog's.
Western Coyote (*Canis latrans*)	15–35 lbs (6.8–15.9 kg)	1½–2⅞ in. (3.8–7.3 cm)	Small track could be confused with any Fox's.
Eastern Coyote (*Canis latrans-lupus or rufus*)	30–60 lbs (13.6–27.2 kg)	2½–3 ½ in. (6.4–8.9 cm)	In most cases, track can be distinguished from Red Fox and Wolf by length.
Red Wolf (*Canis rufus*)	40–80 lbs (18.1–36.3 kg)	3–4¼ in. (7.6–10.8 cm)	Easily confused with Coyote, who may factor heavily in Red Wolf's lineage.
Eastern Timber Wolf (*Canis lupus lycaon*)	50–100 lbs (22.7–45.4 kg)	3½–5½ in. (8.9–14 cm)	Weight and track length highly variable, even in same pack.
Northern Rocky Mountain Wolf (*Canis lupus irremotus*)	70–130 lbs (31.8–59 kg)	3½–6 in. (8.9–15.2 cm)	Size and track can be confused only with Domestic Dog.
Domestic Dog (*Canis lupus familiaris*)	3–175 lbs (1.4–79.4 kg)	1–5 in. (2.5–12.7 cm)	Wide variability, potential confusion with all wild canines including Dog-Coyote-Wolf hybrids.

Over most of the country, Red Fox, Gray Fox, and Coyote coexist, which makes track length overlap an issue. Even more challenging are those areas of the Western U.S. where Coyote and two or even three Fox species ranges overlap. Problems arise in the Northeast and Southeast where large Coyotes coexist with Red and Timber Wolves. The strides, gaits, and track patterns of many canines overlap. If we then add Domestic Dogs, whose tracks can be confused with any of the wild canines, we have the tangle Justin found himself in.

The first step in clearing the fog is to understand why there is such a spread in each species' weight range and track length. Let's use the Eastern Timber Wolf as an example. In Minnesota and Wisconsin, she has an average track length of 4.5 inches (11.4 cm), though it may be a full inch shorter or longer. This is due to several factors:

- Sexual dimorphism. Male Wolves and Kit Foxes average 20 percent heavier than females, while Coyote males range only 10 to 15 percent larger. Gray Foxes show no significant difference.
- Latitude. Known as Bergmann's rule, 80 percent of animal species increase in size the farther north they occur. The Mexican Gray Wolf (*Canis lupus baileyi*) ranges from 60 to 80 pounds (27.2–36.3 kg), while the Arctic Wolf (*Canis lupus arctos*) runs 100 to 125 pounds (45.4–56.7 kg).
- Hybridization. The Eastern U.S. has "Coyotes" as large as and larger than some of its Red and Timber Wolves. Coyote weight increases from West to East, which indicates degree of hybridization. Minnesota's Coyotes average 25 to 35 pounds (11.3–15.9 kg), Wisconsin's range up to 45 pounds (20.4 kg), and New England's reach 60 pounds (27.2 kg), though animals over 50 pounds (22.7 kg.) are rare. This rate of increase is four times that of Bergmann's rule. However, while Coyotes increase in size, Wolves actually grow smaller, with Canada's Algonquin Park Wolves averaging 10 pounds (4.5 kg) less than Minnesota's.

- Maturation Rate. Female Wolves take two years to reach full growth and sexual maturity, and males take three years. This is shown in the yearling-to-adult weight gains in the table below. Coyote females mature in one year and males take an additional year, like Wolf males.

The table, which is based on the weights of 158 recently live-trapped Minnesota Wolves, shows that sexual dimorphism is insignificant in yearlings, with females averaging 61 pounds (27.7 kg) and males 63 pounds (28.6 kg). With adults, however, there is a 17 percent weight difference, and a survey of Wisconsin Wolves shows 19.5 percent.

MINNESOTA TIMBER WOLF MATURATION RATES

AGE AND SEX	NUMBER OF ANIMALS	WEIGHT RANGE	AVERAGE WEIGHT	YEARLING-TO-ADULT GAIN
Yearling Female	20	46–79 lbs (20.9–35.8 kg)	61 lbs (27.7 kg)	—
Adult Female	59	50–90 lbs (22.7–40.8 kg)	64 lbs (29 kg)	5%
Yearling Male	16	43–82 lbs (19.5–37.2 kg)	63 lbs (28.6 kg)	—
Adult Male	63	52–104 lbs (23.6–47.2 kg)	77 lbs (34.9 kg)	18%

Once you get to know your region's wild canines and their tracks, you can make charts to help quickly identify species by their distinguishing characteristics. Based on my study of nearly 1,000 front paw prints, I made figure A2.1 to help quickly differentiate Coyote's track from both Red Fox's and Wolf's in the Yellowstone-Glacier National Park region. I chose a feature unique to a Coyote

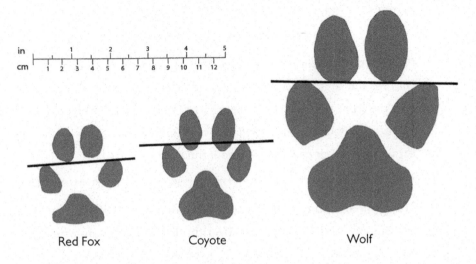

Figure A2.1. Northern Rockies Canine Track Identification

track, which is shown by the line running from tip to tip of the outer toe pads. This line runs through the middle toe pads, while the lines through Fox's and Wolf's tracks do not touch the middle pads. With Fox and Coyote tracks often overlapping in size, this technique helps me identify many of them at a glance. The same is true with Coyote and Wolf tracks, although there is enough of a size difference that I would seldom be confused.

This comparison in figure A2.1 may or may not be applicable to your area, as I have found regional variations in some species' tracks. The method's degree of accuracy in the northern Rockies is around 80 percent for Fox and Coyote tracks and 70 percent for Wolf. It works for Fox and Wolf tracks on all substrates, but only on soft substrates for Coyote tracks. I define a soft substrate as any medium in which paws sink. On hard substrates, Coyote tracks commonly register with middle toes extended, just like Fox and Wolf. (The middle toes of canines in general tend to register shorter on soft substrates and longer on hard substrates.) Fortunately for this method, most tracks with clear-enough detail are found on soft substrates.

Why Not Use the Hind Paw Print?

Most predators have prominent forelimbs with well-developed mus-
culature and large, articulate paws to aid in the chase and kill; there-
fore the front paw print tells the clearest story. Alternatively, prey
animals like Rabbits and Deer have powerful hindquarters for quick
escapes, with rear paws larger than fronts.

Using the method in the Eastern United States is difficult because
of the issue of Wolf-Coyote hybridization. Besides confusing the heck
out of trackers in the Eastern U.S., hybrid populations put to question the protected status of what were once considered to be remnant
or restored Wolf populations. Hybrids may not qualify for protection,
especially those with high percentages of Coyote blood. In most areas,
year-round Coyote harvesting is allowed, and already hunters have
trouble distinguishing large Coyotes from small Wolves.

Genetic studies are starting to provide answers for both trackers and
wildlife managers. But first, let's look at the problem. Figure A2.2 shows
simplified versions of two popular Wolf-Coyote evolution models.

Model 1 presents the Gray Wolf, Red Wolf, and Coyote as distinct

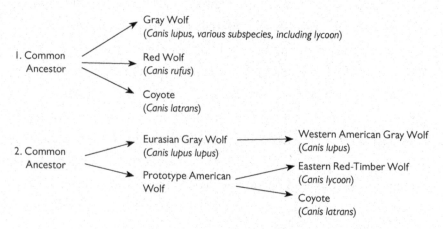

Figure A2.2. Wolf-Coyote Evolutionary Relationship

and parallel species. Conversely, model 2 has the Eastern Wolf more closely related to Coyote than to the Western Wolf. This relationship is used to explain why Eastern Wolves hybridize with Coyotes and Western Wolves do not.

A third model is starting to emerge from the latest genetic research. It shows only two gene pools—Wolf and Coyote—with the Western Great Lakes Wolf, Algonquin Park Wolf, Red Wolf, and Eastern Coyote all being Wolf-Coyote hybrids to varying degrees. Figure A2.3 gives a graphed depiction.

If further testing confirms these findings, the Red Wolf may have to be acknowledged as a Coyote with a minor amount of Wolf blood, similar to the Northeastern Coyote. Ironically, some Northeastern Coyotes were found to have more Wolf blood than most Red Wolves

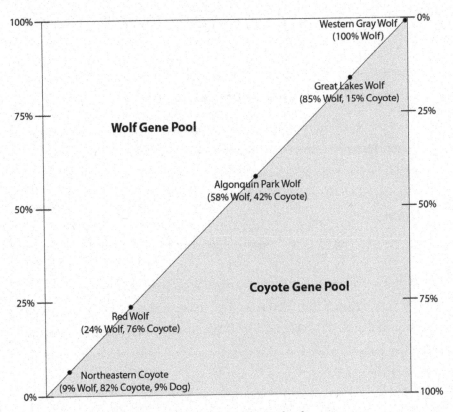

Figure A2.3. Wolf-Coyote Hybridization

and many Algonquin Park Wolves. With nearly half Coyote blood, the Algonquin Park Wolf might be more properly called a Coy-wolf.

Let's keep in mind that the percentages in figure A2.3 are averages. Members of hybrid populations usually have varying percentages of each parent stock's blood. One animal could fall strongly on the Wolf side of the genetic spectrum, while the next one will be mostly Coyote. Even with the same genetic composition, as with littermates, individuals can appear to be either more Coyote-like or Wolf-like. These variations were clearly evident to me, after living for five years with a pack of Western Great Lakes Wolves back in the 1970s. The alpha female weighed around 50 pounds (22.7 kg) and had the petite build, refined muzzle, and pointed ears of a Coyote. Her sibling, the alpha male, ran 75 pounds (34 kg) and had mostly Wolf-looking features. And then there was the 100 pound (45.4 kg) brother, who sported the robust build, squared muzzle, and rounded ears that made him look Wolf to the core.

Like Wolf, Like Fox

In West Texas there is evidence of Red Foxes hybridizing with Kit Foxes. European Red Fox subspecies were introduced to America in the eighteenth century and interbred with indigenous American populations. Foxes cannot hybridize with Coyotes, Dogs, or Wolves.

Genetic evidence shows two Coyote-Wolf hybridization events: 600 to 900 years ago in the Western Great Lakes area and 50 years ago in Algonquin Park. The Algonquin Wolves will likely continue sorting out their genetic ancestry until they come up with the right mix for their environment.

Notice that figure A2.3 lists the Northeastern Coyote's genetic makeup as 9 percent Dog. Many Wolves also carry Domestic Dog blood. The genetic mutation that causes melanistic (black-pigmented) Wolves originated with Dogs, and black Wolves are common across Southern Canada and Minnesota. Around half of Yellowstone's Wolves are black. The melanistic gene also occurs in Red Wolf and Coyote populations.

Considering this, it would seem we need to add a third gene pool to figure A2.3. Prior to 1993, we would have had to, as the Domestic Dog was classified as *Canis familiaris*, a distinct species. However, in that year the Dog was reclassified as *Canis lupus familiaris*—a Wolf subspecies. The bulk of scientific evidence points to all Dogs being descended from Wolves, and only 15,000 to 40,000 years ago, which is quite recent in evolutionary terms.

It is hard to imagine a little shag-carpeted lapdog or a St. Bernard being a Wolf, but all Dogs have some un-wolf-like physical characteristics and mannerisms. Here is why:

- Dogs are highly specialized Wolves. Selective breeding has magnified certain Wolf traits, such as the scent-tracking ability in Bloodhounds, which is ten times that of Wolves.
- Dogs continue the mutant genes causing such characteristics as short legs and compressed faces.
- Many Wolf traits are diminished in Dogs. A Bloodhound is much more the scent tracker than a Wolf, but he is not near the runner.

There is evidence of Wolves interbreeding with Dogs in most areas of the world where the two coexist, and it has probably been occurring all along. When a Wolf population is infused with Dog genes, those which hamper survival in a Wolf's world are eliminated. Occasionally a gene mutation is retained, as with forest-dwelling Wolves where the melanistic gene gives an adaptive advantage (possible protective coloration and disease resistance).

This knowledge of canine genetics and hybridization has helped me considerably with tracking and species identification. I no longer feel compelled to micro-analyze tracks in areas where there is a mishmash of Wolf-Coyote, Wolf-Dog, and three-way crosses. Wolves with Domestic Dog blood are now simply Wolves. Timber Wolf, Red Wolf, and Eastern Coyote are members of the same hybrid family, differing only in which side of their ancestry they lean toward.

Distinguishing wild canine tracks from those of Domestic Dogs can be every bit as challenging as telling one wild canine track from another. There are Dogs whose tracks and mannerisms could be confused with those of any wild canine on the continent. Here are some puzzle solvers:

- Use the guidelines set forth in tracking books by Elbroch, Halfpenny, Lowery, Moskowitz, Murie, and Rezendes.
- Familiarize yourself with the ranges, habitats, and behavioral characteristics of your area's wild canines.
- Use the rule of three: find three unrelated pieces of evidence that corroborate each other.

I hope the information and techniques presented here will help develop your track recognition skills. If you are a novice, remember that the more you practice, the better you will become. And the process will help waken your intuitive tracking ability—it is not unreasonable to imagine yourself someday identifying most tracks at a glance. Along the way, you'll have the joy of coming to know the living track—and you will pick up some good stories to tell.

APPENDIX 3

Tracking at a Glance

A Symbol Reading Primer

In chapter 14, "Following the Flight Trail," Phil and Antoine so easily identified their Hawk not because of anything I or anyone else taught them, but because of their innate ability to identify symbols. It is a good example of the one reason we evolved this skill to such a high degree—to help us find our quarry as quickly and reliably as possible. Unfortunately, symbol tracking is only lightly covered in the tracking literature, so here I will give some basic information.

Track and sign details give us precise and reliable information; however, it may be more than we need in order to stay on the trail. Additionally, we have to stop periodically to read it. With symbols, we can often get the information we need at a glance and not have to break stride.

Symbols are easily identifiable shapes that represent something more than the shape itself. An octagonal stop sign and a raised fist are symbols. We are continually—and instantly—interpreting symbols: words, facial expressions, cloud formations, and so much more. The reason we can interpret them in an instant is that they do not need to be consciously recognized. Right now you are interpreting words, a common symbol for thoughts and feelings, without giving the process any thought.

Symbols on the trail work in the same way that glances at people's faces instantly identify them. I don't need to recall the colors of their eyes and the shapes of their noses, because I already have their symbolic image imprinted in my mind. All it takes is a quick glance for me to match face with symbol. It is much faster and more accurate than taking measurements and comparing ratios to identify people. As with animal tracks, more than one person could fit the dimensions. Phil and Antoine closed their case right away when they saw the symbol.

Most symbols are embedded in what they represent, like the X formed between the toes and metacarpal (palm) pads of a canine (see figure A3.8, page 267). Other symbols are only suggested by the track, such as the trapezoid formed by the tracks in figure A3.6 (page 264).

Many modern trackers tend to literally read the track—they see it as a book to be opened. Reading, however, takes time. One reason aboriginal trackers can trail so fast is that a track's symbol makes an impression upon the mind that can be "read" instantly. This is possible because the mind relates to the world geometrically. When we look at the stars, we see relationship patterns that are not just stars. Some people call this special relationship we have with the cosmos *sacred geometry*. Without it, the stars would be no more than a bunch of bright dots. With it, the stars speak. And so does the track.

Sometimes toe and pad impressions—even the track outline—are hard to distinguish, yet the track symbol remains visible. Such was the case this morning when I came upon a patch of dry, sugary sand. Cutting across the far corner was what looked like a couple of tracks. Figuring they'd be badly distorted, I decided to have a little fun by first giving them a good analytical look. The sand was so soft that it washed back into the tracks, leaving little more than round depressions. From the size and stride, the animal could have been a large house Cat, a small Fox, or a female Fisher. The straddle seemed wide for a Fox or

Cat, but then I recalled that I myself widen my straddle on soft or slippery substrates. And there were only two prints—not enough to make a good determination of stride or straddle. One of the prints looked as though it could have been an indirect register, which was not typical of Cat or Fox on soft substrates. But then this was sugar sand, which is slippery as well as soft. I wondered what other animal this could be.

And so went my analyzing, until I felt I had done the rational approach justice. I then walked away from the sand patch, centered myself, and reapproached. Walking by the patch, I just glanced at the tracks and there it was—a faint sand pyramid where the two bars of the X, a Canine track symbol, crossed. Nothing else was distinguishable but this symbol remnant—which I could pick up when not looking too hard.

How Symbols Work

Some symbols, such as facial expressions, are probably imprinted in our genetic memory. A baby seems to automatically know the meaning of a smile, a frown, a look of surprise. We likely inherited symbol forms such as these from our pre-human ancestors. I've witnessed the same ability in young birds and canines I have raised in isolation from their own kind. This is not surprising considering that we have the same core brain as birds and mammals, which we all inherited from our common reptilian ancestor. This brain takes care of self-preservation, reproduction, and basic emotional responses. It gives us the capacity to become and shadow the animal we are tracking, and it houses our symbol library.

We appear to inherit the symbol imprint, and thus inherently recognize, at least one family of animals—Snakes. People across cultures fear them, even with no prior exposure, and often refer to them as slimy, when they are in fact dry and scaly. One explanatory theory suggests that at some stage in our evolutionary past, we lived on the shores of a shallow sea where biting Eels were common. We had to react quickly to the sight of an Eel, so we imprinted on her form in order to respond instinctively. Having a form similar to Eels, Snakes get the same reaction.

Another inherited symbol seems to be the shadow. Many animals, us included, instinctively react in fright to an unexpected shadow, especially if moving. Some people affix Hawk silhouettes to windows to deter small birds from crashing into them. It is commonly believed that the birds react to the silhouettes because they are Hawk-shaped. In actuality, a variety of shapes will work, as long as they are Hawk size, dark to simulate a shadow, and have some movement. A decoy or picture of a Hawk is not as effective as a silhouette. A picture has to be processed, which takes time, and reaction to a symbol is instantaneous. Verbalizations have to be processed as well. A raised fist gets an instant reaction, whereas angry words need time for comprehension.

The first recorded verbalization was in symbols, commonly known as pictographs. They were universal symbols that told stories each reader could interpret in a way that was meaningful to him. On the other hand, our contemporary written stories require a level of literacy, which is essentially memorizing and utilizing new symbols.

We can add symbols to our library because of the memory and temperamental capacities of our second, or mammalian, brain. One of the first symbols infants add to their repertoires is their mothers' facial features. Newly hatched Ducklings imprint on a mother as well, only it does not have to be a Duck. Dogs, Cats, Owls, humans, even toys, have filled the niche.

With our primate brain, we humans can go another step beyond the capacity of our mammalian brain and engage in the sophisticated communication, planning, and envisioning that make us the remarkable trackers we are. Equally incredible is the ability our primate brain gives us to do so many things with our prey other than just filling our bellies, such as using the inedible parts to fashion tools, clothing, shelter, and musical instruments.

Symbol Tracking Tips

If we do not use our symbol tracking ability, we disregard a part of what makes us human. And we compromise our tracking. Remember that when we can identify and interpret tracks at a glance, rather than having to stop and read track and sign, we can keep moving rapidly on the trail. When tracking is a hobby, take your time. When tracking to feed yourself, speed and efficiency are essential.

Not So Fast

When symbols give easy information at a glance, novices are often tempted to run with it without giving it a second thought. This may be necessary on the chase or in response to a shadow, but at other times, remember to be as a question.

To get into the habit of keying in on the symbol, ignore the track's details. They are usually not clear anyway, as most trails cross poor tracking substrates and details are the first part of the track to erode. Look beyond them into the soul of the track—the symbol—which usually outlives the track's body.

As you'll see in the following illustrations, most symbols look nothing like the track. The rational mind usually discounts them, so it is important to go with your first impression, your intuitive voice. When it feels right—not when it makes sense—you can be reasonably sure that you've clicked on the symbol and you can go with it. The symbol-track relationship is the same as that with another person: when it is right, it just feels right, no matter what the mind says.

Think of a track symbol as a pictograph. Symbol tracking is allowing the pictograph within the track to connect with the pictograph already in our mind. Notice I said "allowing." Because the symbols are already there, our only work is to get out of the way. We can help this process by re-enlivening the part of the reptilian brain that holds the symbols. Here are two ways:

- Draw the track on the ground right beside the actual track. The process can open neural channels and help imprint the track symbol.
- Visually relate track to symbol over and over until it starts to automatically click. Following are examples of several symbol types for you to use in the process.

Trying Too Hard

Sometimes the symbol is so obvious that we overlook it—especially when we are intensely focused on the track. We are not allowing the mind to do what it does best, which is to observe and free-associate. Instead we force it to analyze, which overshadows what we would intuitively be drawn to.

Silhouettes

For small prey animals, reacting to silhouettes is a way of life. So many generations of small birds have been harassed by raptors that all it takes is a dark shadow passing overhead to trigger their fight-flight mechanism. A raptor perched on a branch will not bring anywhere near the response of her moving shadow. The response is so instantaneous that the shadow symbol must be imprinted in their genetic memory. My observation suggests that the shadow doesn't have to be in the form of a raptor; it only needs to be moving and the approximate size of a raptor.

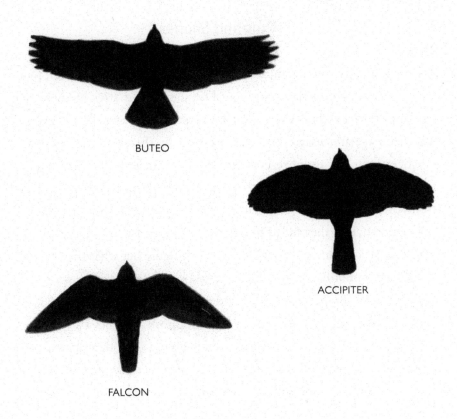

BUTEO

ACCIPITER

FALCON

Figure A3.1. Raptor Silhouettes

Still, raptor family silhouettes can be helpful identification aids for us. There are times when the silhouette—or just a part of it—is all that we catch when a bird flashes across our field of vision. With the help of silhouettes, the shape of the tail or the width of the wing may be all we need to identify the bird. As with any sign, knowledge of habits, habitat, and range work hand-in-hand with silhouettes to help identify the bird. Above are the silhouettes of three common raptor families: Buteo, Accipiter, and Falcon. Familiarize yourself with them until you can recognize them instantaneously. Then have someone quiz you by covering up all but one feature, such as the tail or tip of a wing.

Forms and Shapes

Dens

Most small and some medium-sized mammals use dens for shelter, food storage, and reproductive purposes. The shapes and sizes of their den openings form symbols that can help identify their owners at a glance. The den's age and usage can be determined by the excavated material's rate of erosion, the age of vegetation growing on it, and other sign such as fur and food scraps. The students in chapter six, "How to Learn Tracking from One of the Greatest Predators," give a good den-aging demonstration, even though they failed to notice a couple of the things mentioned here. Go back to the story and see if you can figure out what they missed.

Two commonly confused dens are those of Badger (*Taxidea taxus*) and Red Fox (*Vulpes vulpes*). Rick describes a Badger den in chapter 9, "Grandfather Tip-Up." Figure A3.2 (opposite) shows the symbols for the den entrances, followed by a table giving their descriptions and other distinguishing characteristics. Den entrance size and shape will vary according to slope, soil type, and the individual animal.

Lays

A quick look at an animal's bed, which is typically called a lay, can give a wealth of information. Its location, size, and symbol identify the species, and its condition tells when it was used. Often it is possible to identify the specific animal and why she used the lay. It could have been for rest, shelter, hiding, nursing, or waiting in ambush. Lays are typically found in locations that afford safety and a good lookout. They also potentially provide warmth, shelter, or escape from biting Insects.

Lays are most easily found in snow, soft vegetation, and sand, though other substrates can hold lay imprints as well. On page 258 are the symbols of a deer lay and a canine lay found in snow. Both are typical for the family and will vary in size according to species and age. Winter Whitetail Deer lays in my area range from around 30–43 in. (75–110 cm). Envision the laying animal with the help of the labeled body parts.

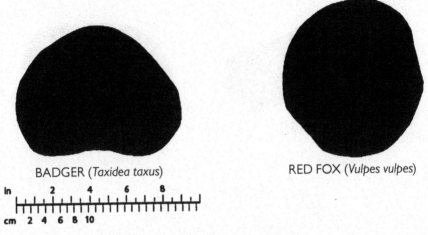

BADGER (*Taxidea taxus*)   RED FOX (*Vulpes vulpes*)

Figure A3.2. Den Openings

SPECIES	DIMENSIONS	UNIQUE FEATURES	CAN BE CONFUSED WITH . . .
Badger	Elliptical, avg. 10 in. (25 cm) wide and 8 in. (20 cm) high.	Often widest at base, hump in the middle of opening, excavated material spread in a fan from breaststroke digging style.	Woodchuck's (*Marmota monax*), in the Upper Midwest and West-central Canada, where their ranges overlap. Badger dens often have claw marks on sidewalls. Woodchuck front entrances have high mounds and escape holes have none.
Red Fox	Oval, avg. 9 in. (22 cm) in diameter, often taller than wide.	Highly variable shape, sometimes pinched near base.	Coyote's (*Canis latrans*), whose den opening is similar, but ranges from half again to over twice the diameter. Wolf's is half again the size of Coyote's.

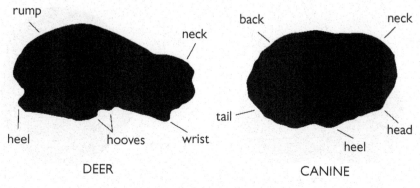

Figure A3.3. Lays

With practice on a variety of lays, you should start seeing the symbol at a glance, and then the animal within the symbol.

Gnaw Marks

The incisors—the large, paired, front cutting teeth—of rodents and lago-morphs (Rabbit family) leave distinctive gnaw marks that can be quickly identified if you are familiar with them and the related sign. Besides gnawing on plant matter, bones, and antlers to obtain food, these animals must gnaw to keep their constantly growing incisors worn down.

Figure A3.4 shows gnaw-mark symbols for six animals common to my area.* Add the symbols of your local gnawers not depicted here and quiz yourself until you can identify them at a glance. Then have some-one quiz you by covering all but a small portion of the symbol.

Gnaw marks often overlap or only partially register, and the deeper they are, the wider they appear. You will not often find them as well-defined as they are here. Knowledge of food and habitat preferences, along with scat and tracks, will give additional clues to identifying the animal.

Members of the Deer and Goat families eat tree bark; however, they lack the upper incisors to make clean cuts, so they scrape and tear more than gnaw. They work from the bottom up, so look for the telltale

---

*All data and illustrations are from the Lake Superior bioregion. Data from other areas will reflect their variant populations.

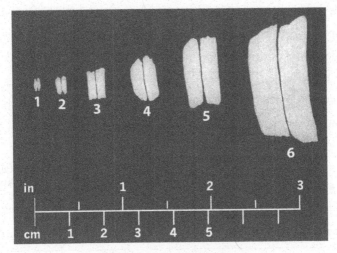

Figure A3.4. Gnaw Marks

1. Deer Mouse (*Peromyscus maniculatus*)
2. Red Squirrel (*Tamiasciurus hudsonicus*)
3. Snowshoe Hare (*Lepus americanus*)
4. Muskrat (*Ondatra zibethicus*)
5. Porcupine (*Erethizon dorsata*)
6. Beaver (*Castor canadensis*)

ragged edge at the top. A mature Whitetail Deer in my area can have middle incisors that are each ½ in. (13 mm) wide, but it is usually the smaller side incisors that register.

Rodents' gnaw marks usually run perpendicular to the branch or trunk, while those of Deer run parallel. A Deer's gnaws look haphazard next to a rodent's orderly rows. Rodents cut branches cleanly off at an angle, while the Goats and Deer tear them off, leaving a ragged edge. The height of a gnaw mark may not be a reliable species indicator, as tall animals will gnaw at the base of a tree and short animals are hoisted by deep snow.

The table on the next page gives additional information to help identify these gnaw marks in the field. To augment this, collect your own data, which will reflect the species variation in your area. Take advantage of others' research, and add species endemic to your area.

Keep in mind that data is only a guideline, as sign is often fragmented and impossible to measure. Just as with humans, there is wide variation in the dentition of any particular animal species. You will find wide and narrow incisors, with some spaced and some tight together. Lower incisors are generally narrower than uppers, yet I found a Snowshoe Hare with lowers outshining his uppers by 20 percent.

SPECIES	AVG. UPPER PAIRED INCISOR WIDTH	UPPER-LOWER DIFFERENTIAL	SEXUAL DIMORPHISM REFLECTED IN INCISORS
Deer Mouse	$\frac{1}{16}$ in. (1.2 mm)	-10%	Negligible
Red Squirrel	$\frac{1}{8}$ in. (2.5 mm)	-15%	Negligible
Snowshoe Hare	$\frac{3}{16}$ in. (5 mm)	-4%	Negligible
Muskrat	$\frac{1}{4}$ in. (6.5 mm)	-12%	Negligible
Porcupine	$\frac{5}{16}$ in. (8 mm)	Negligible	Negligible
Beaver	$\frac{9}{16}$ in. (15 mm)	Negligible	Negligible

Note that the width difference between upper and lower incisors is most pronounced in small rodents, yet it is usually not significant enough to cause one species' measurements to fall within the parameters of another. Sexing the animals listed is not possible by the width of gnaw marks, as they are nearly identical for males and females.

Experience can gradually take the place of your ruler, and you will be able to identify gnaw marks by sight.

Go Snow

"Look ever for the track in the snow; it is the priceless, unimpeachable record of the creature's life and thought, in the oldest writing known on the earth." —Ernest Thompson Seton

---

Symbol tracking is most easily learned in snow, which is where we found all of the tracks and symbols in the following section. Note that all track outlines are taken at the original snow surface, ignoring any upheaved snow.

Track Patterns

Dumbbells

The symbol created by an animal walking through deep snow often resembles a dumbbell. The foot cuts down into the snow, the narrow ankle slices forward, and the foot lifts out, completing the dumbbell. Figure A3.5 (on the next page) shows four different Whitetail Deer dumbbells.

Dumbbells show:

- **Direction of travel.** In Dumbbell 1, imagine water running off of the plateau onto the plain and you have the direction. The left end of Dumbbell 3 shows the flat bottoms of the front and rear hooves sliding into the snow, while the right end registers the rounded wrist pulling out.
- **Age.** Dumbbell 1 was made before the snow iced over and Dumbbell 2 was made after. Dumbbell 4 can be aged by the rate of snowmelt. A dumbbell's crisp edges gradually round out from snow evaporation. Learn the rate of evaporation by observing the effects of wind, temperature, and humidity over time.
- **Right or left foot.** The saddle is a reliable indicator of either the inside or outside of the dumbbell, and I will leave it for you to discover which it is.

After little practice, you should be able to read a dumbbell at a glance. Note that in all four illustrations—even highly degraded 4—the plateau, plain, and saddle are recognizable.

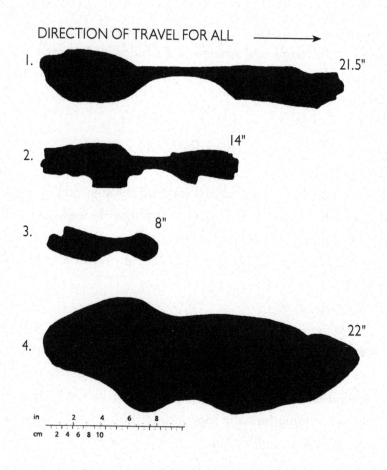

Figure A3.5. *Whitetail Deer Dumbells*

1. A long step in soft snow
2. A step in ice-crusted snow (notice the sharp angles)
3. A short step
4. An old, melted-back step

Trapezoids

Deep snow is a great equalizer for the Northcountry's animals. The large ones might have a graceful walk, tireless lope, or speedy gallop, but shoulder-deep powder can force them to join their small kin and hop. An animal hops by landing on her front feet, bringing her back feet up around them, and pushing off with back feet to become airborne.

Always the Exception

Moose is able to move more easily through deep snow than most other large animals, due to his great weight, stilted legs, and the capacity for his rear legs to lift out of the snow the same way they entered. Porcupine seems oblivious to the snow—she waddles along the same as always, only plowing a trough as she goes.

I once came across the tracks of a Coyote chasing a Deer through a mixed conifer-hardwood forest. The knee-deep snow wasn't enough to break the Deer's gallop, but it slowed him up enough that Coyote was gaining. The Deer took an abrupt turn for the adjacent open bog, only the snow there was shoulder deep. No problem—he flew over it in a series of great twenty-foot hops. Coyote, with his short legs and light weight, got bogged down in the fluff. On the other side of the bog, the Deer slowed right down to a walk, knowing Coyote was floundering out there somewhere.

Small rodents and lagomorphs are designed to hop, and they aren't heavy enough to sink out of sight like Coyote, so they do quite well in deep snow. Figure A3.6 (on the next page) shows the track outlines of six common Northwoods animals in shallow snow, deep snow, and powder. Remember that these outlines appear at the snow's surface.

Notice how the deep snow tracks resemble an open or closed-bottomed V. They are formed by the heel of the back foot sinking in and merging with the front footprint to create troughs. Powder tracks, which look like a ♡, sometimes with a bottom cleft, are made by the body print merging with the troughs. Combine this information with knowledge of local weather and you can age tracks. A ♡ was laid soon after a fresh snowfall; a V came after the snow firmed up a bit, and individual pawprints register on hardpack snow or on a thin layer of new snow over old crust.

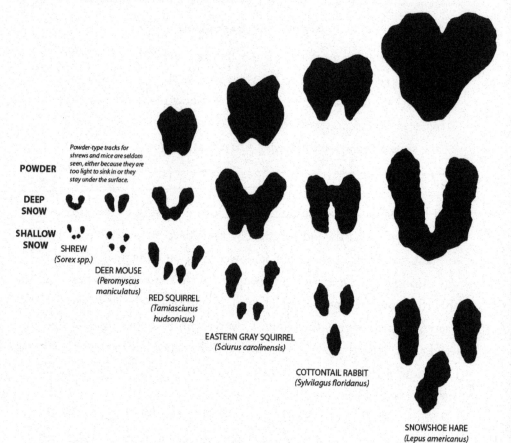

POWDER

Powder-type tracks for shrews and mice are seldom seen, either because they are too light to sink in or they stay under the surface.

DEEP SNOW

SHALLOW SNOW

SHREW
(Sorex spp.)

DEER MOUSE
(Peromyscus maniculatus)

RED SQUIRREL
(Tamiasciurus hudsonicus)

EASTERN GRAY SQUIRREL
(Sciurus carolinensis)

COTTONTAIL RABBIT
(Sylvilagus floridanus)

SNOWSHOE HARE
(Lepus americanus)

Figure A3.6. Trapezoids

The common symbol for all of these tracks is the trapezoid: ▽. Knowing this, a glance at a snow track can tell you whether an animal falls into this grouping, and if so, the following information will give you the species. Mouse's trapezoid is the standard shape, whereas Shrew's is squat: ▽. The same difference can be seen between the Squirrels. They can be quickly differentiated from the lagomorphs by looking at the placement of the front feet: Squirrel's will be side-by-side, while Rabbit and Hare generally place one ahead of the other (which is typical of ground dwellers). Shrew's V is generally scrunched, Mouse's is often split, Squirrel's can be widely splayed, and the arms of a lagomorph's tend to be close to parallel. The two Squirrels have free-form ♡ tracks, which differ markedly in size, as do the better-formed Cottontail's and Snowshoe's (see figure A3.6).

Use Life-Size Illustrations

To learn tracking at a glance, enlarge track illustrations to life size. Their symbols will imprint in your memory for quick field identification, as the mind will not need to process them to account for the size difference.

Although not entirely reliable because of overlap, trail width—the average outside spread of the rear foot tracks at the snow's surface—can be an adjunct to species identification. The following table gives the general range of trail widths in my neck of the woods.

SPECIES	TRAIL WIDTH, SHALLOW SNOW	TRAIL WIDTH, DEEP SNOW
Shrew various species	¾–1¼ in. (1.9–3.2 cm)	Same
Deer Mouse	⅞–1⅝ in. (2.2–4.1 cm)	Same
Red Squirrel	2½–3½ in. (6.5–9 cm)	2½–3¾ in. (6.5–9.5 cm)
Gray Squirrel	2¾–4 in. (7–10 cm)	3½–5 in. (8–13 cm)
Cottontail Rabbit	2½–4½ in. (6.5–11.5 cm)	3½–5½ in. (8–14 cm)
Snowshoe Hare	4–6½ in. (10–16.5 cm)	6–7½ in. (15–19 cm)

In powder, a Snowshoe's trail can be as wide as 8½ in. (21.5 cm). To learn your area's trail widths, take your own measurements. A minimum of ten are needed to get a reliable average. Consider using your hand instead of a ruler. In time you will need just your eye.

Where one person sees trapezoids and Vs, another person sees butterflies and hearts. Here is a trail for romantics that one of our staff members photographed (figure A3.7). The shape of the top track gives clues as to how a heart is formed.

Figure A3.7. String of Hearts

Hoops and Hexagons

When Dogs, Coyotes, and Foxes traverse the same territory as Domestic Cats and Bobcats, novice trackers can have trouble telling canine tracks from feline. With both having four toes and a palm pad, and often being around the same size, there is bound to be some confusion. At the same time, they have differences that allow experienced trackers to distinguish them fairly easily.

There is a common symbol found in feline front foot tracks, and another one found in canine front tracks. Whether we are beginners or experts, these two symbols can quickly do the identification work for us. The Cat family's is ◯ and the Dog family's is ⟨⟩. They will vary somewhat with species, and with the factors at play when the track was laid. Sometimes the symbols are apparent even in degraded or partial tracks. To see the symbol, step back and let the track speak. Try it on the tracks illustrated in figure A3.8 (below). If the symbol is not apparent, draw it around the track to help you make the association.

Most people have no trouble seeing the circle symbol in Cat family tracks, but confusion arises with the spread-toed front tracks of Gray Foxes and many Domestic Dogs. In fact, any canine can leave a circular-looking track when spreading toes to gain support and traction on a soft substrate. When you are not sure of the track, step back. If it is a canine's, an equilateral hexagon, ⬡, formed by the four toes and the two lower corners of the palm pad, should appear.

In the Coyote track in figure A3.8, notice that one outer toe sits higher than the other. This characteristic is a right-left foot indicator, and it applies to Eastern Timber Wolves as well. You can learn how to use it by observing tracks and track photos. Bobcat and Cougar tracks show a similar feature—in the Bobcat track in figure A3.8, notice the drift of the toes to one side.

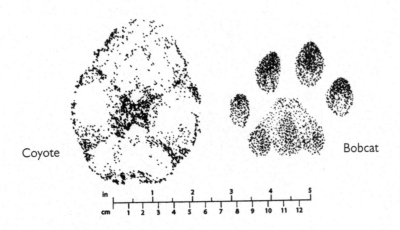

Coyote   Bobcat

Figure A3.8. Hoops and Hexagons

Kill Site Patterns

The best whodunits I've ever read were written in feathers, fur, bones, and blood. When I come across one of these stories in the woods, I'll sometimes stop and read every page, and at other times I'll get the story from a glance. In those cases, I've caught a symbol that connects with a story I already know. When the prey is a bird, the symbol is embedded in the predator's dinner table feather arrangement. Figure A3.9 shows the most common feather configurations, their symbols, and the related stories.

Naming the symbols could help you remember them. I call them: 1. Wreath ◎, 2. Circle ◯, 3. Egg ◌, 4. Jumble ⸜⸝. Or if you have trouble remembering anything but your next snack, try: 1. Donut, 2. Cookie, 3. Chocolate egg, 4. Honey nut granola.

In figure A3.10, numbered references 1 and 4 may be either kill

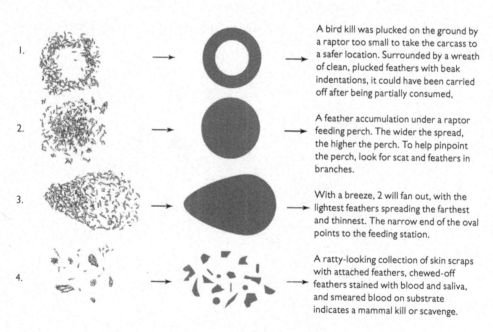

1. A bird kill was plucked on the ground by a raptor too small to take the carcass to a safer location. Surrounded by a wreath of clean, plucked feathers with beak indentations, it could have been carried off after being partially consumed,

2. A feather accumulation under a raptor feeding perch. The wider the spread, the higher the perch. To help pinpoint the perch, look for scat and feathers in branches.

3. With a breeze, 2 will fan out, with the lightest feathers spreading the farthest and thinnest. The narrow end of the oval points to the feeding station.

4. A ratty-looking collection of skin scraps with attached feathers, chewed-off feathers stained with blood and saliva, and smeared blood on substrate indicates a mammal kill or scavenge.

Figure A3.9. Feather Talk

sites or feeding sites, as predators often move kills to safe locations for consumption. You may be able to tell which it is by the pattern of the accompanying blood. When smooth substrates like snow and rock leave the blood stain clearly visible, its symbol can jump out at you. On broken groundcovers like grass and forest duff, the stain's form may not at first be apparent. Change your viewing angle to see through the surface layer and get a three-dimensional perspective. Look for telltale variations in color and sheen.

Figure A3.10 (below) shows the four classic symbols found in stains: 1. Blotches, 2. Spots, 3. Streak, and 4. Drip line.

Like feather-grouping symbols, each blood symbol tells the story of its relationship to the kill. The table on the following page gives the start to the story, and your experience will flesh it out.

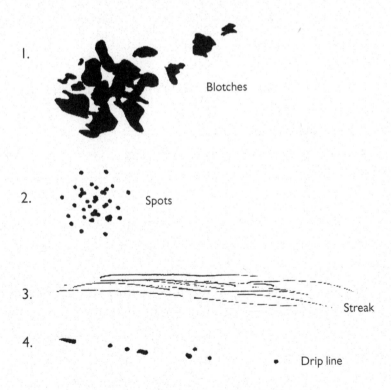

Figure A3.10. Cold Blood, Hot Trail

SYMBOL	STORY OUTLINE	CHECK
Blotches	Multiple blotches indicate an animal was killed, butchered, and/or eaten on site.	For streak or drip line. If it is incoming, the animal was killed or dismembered elsewhere.
Spots	Target-pattern spots say a kill was taken to a perch directly above by a raptor or climbing mammal.	Tree trunk for blood, which could indicate the animal was carried up rather than flown.
Streak	A kill, possibly partially eaten and too large to be carried, was dragged from one spot to another.	Disturbance from drag for direction of travel. Backtracking will take you to kill site.
Drip line	A kill or portion thereof, small enough to be lifted, was carried from one site to another.	Additional sign to see if blood is from a carcass or a wounded animal.

The most common mistakes in interpreting "kill sites" are:

- assuming they are kill sites,
- assuming scavengers are predators, and
- assuming that animals who died of other causes are predator kills.

Remember that anything and everything following an assumption could be wrong, and that the most successful trackers ask the most questions.

There is more to read in kill sites—and all sites for that matter—than what a glance and a symbol can give. When you have the time, leave everything behind and become the site. Take what the symbol gives you, read the sign, and listen to the song of the track. Become the predator—feel the hunger, the keenness of your senses. Then become the prey—alert and suspicious, driven by hunger. The two forces meet and you find yourself enmeshed in an unfolding life-and-death drama. You are now a tracker.

Now Go Back . . .

and revisit the kill site in chapter seven, "A Winter Riddle." You may see something you missed the first time. And be on the alert if it's too quiet for comfort, or you might be the next one to have your feathers arranged in a comely wreath.

How to Act around Bears

As our closest relative, Bear can teach us much about tracking humans, which I relate in chapter fifteen, "Human Tracking, Bear Style." While it is fortunate for our training that Bear lives near so many of us and can be closely followed, I would be acting irresponsibly as a teacher and wilderness guide if I were not to include what she taught me about doing it safely. If you do not abide by these guidelines and caveats and get adequate training, you will put your life at risk.

For safety reasons, Black Bear is the only North American species you should consider getting trained to follow. Start by learning about her character, habits, and habitat preferences. Ranging over 50 percent of the continent, she can be seen at any time or season, especially in the south where food is available year-round. Even in the north, a hibernating Bear will come out to stretch on a warm, white season afternoon. They're very active from mid-green season to the first snows, when they pack in all the food they can find to put on winter fat. During this time, the chance of Bear encounters for woods wanderers is high. It is even greater when Bears get agitated during hunting season and when hunters train Dogs by having them chase Bears.

The first thing I learned is that much of what is popularly believed about Bear disposition, which ranges from fatherly Smokey the Bear to dimwitted bruin, is myth. To know her for who she actually is may be the best thing we can do to act wisely and travel safely in her domain.

Perhaps the most important thing to recognize about the Black Bear is that she is anything but half blind and lumbering. When lean, she can run thirty miles an hour and zip up a Tree as fast as a Squirrel. Her broad, padded feet allow her to move very quietly when she wants to. Her hearing, though I find it to be not as sharp as Wolf's or Deer's, is better than yours or mine. She is not as near-sighted as many think; she can usually see me at one hundred yards. At close range she relies equally on sight and smell. However, unlike her sight, her sense of smell is extraordinary. Her long muzzle says this—the longer an animal's is, usually the better the sense of smell, because of the space it provides for olfactory sensors.

As far as being a threat to us, she is nearly all bluff. When she resorts to loud whoofs and teeth clacking, it is out of nervousness and fear rather than aggression. Even when she slaps the ground and lunges, she is just bluff-charging. She is telling you, "Back off," because she feels crowded. Keep in mind that we're talking Black Bear here, not her cousin the Grizzly. He is entirely another story—don't even *think* about using these tactics with him.

And then there is the most popular myth—that a sow will attack someone who gets between her and her cubs. No matter how big an upset display she puts on, and no matter how loud the cub bawls, it is highly unlikely that she will attack. For the sake of her other cubs, she will not endanger herself. Again, this is Black Bear, not Grizzly. I know of researchers who handle even wailing cubs with no problem, yet please don't take this to mean I recommend doing it.

So is there anything to fear about Black Bear? Oh yes. Silence! Almost all killings of humans have been not by campground beggars but by true hunters who have silently stalked and pounced upon their prey. Bears *are* predators. *Remember: visible and ornery = safe; quiet and stealthy = dangerous.* I know this not only contradicts common knowledge, but common sense as well. However, if we're going to be safe in Bear country, we have to let go of our misconceptions and listen to what Bear tells us.

Playing dead when attacked is another piece of bad advice that is commonly given. Dead prey is exactly what a predator wants, so why try to look like a ready-to-eat meal? If you want to live, you need to out-Bear the Bear. Get up and puff up. Growl, kick, punch, club with a stick, bash with a rock—anything to show you're bigger and badder than she is. Above all, DO NOT RUN. It is a sign that a predator reads as "I'm prey, I'm vulnerable, come and get me." There is nothing a predator relishes more than the chase.

Some people think they're safe climbing a Tree. Unless you're faster than a Squirrel, I wouldn't try it. You're only going to tempt the Bear to scramble right up after you. Still, we have an advantage over her. She may be the most intelligent four-legged predator, but most humans are smarter. If we think like a Bear, we should be able to outwit and even out-act this master bluffer.

Bear's habits can vary from region to region. For local hiking and camping guidelines, consult a seasoned tracker or forest ranger.

APPENDIX 5

Safety Tips in Cougar Country

As with Black Bear in appendix 4, I strongly recommend that anyone venturing into Cougar territory be well-briefed on safety guidelines. While I would be remiss in not covering the topic, and while Cougar attacks on humans have been escalating in recent years, I want to lend perspective by stating that the issue is blown out of proportion by media hoopla. Cougar-caused deaths average one per year, and it is all over the news. At the same time, you're lucky to hear anything about the twenty-five or so people killed by Dogs, the hundred by Bees, or the even higher number by Deer. These much deadlier but well-liked animals just don't make good press.

Cougar has been spreading eastward from the mountain states to re-inhabit his former range, which includes all of the lower forty-eight states. In the past fifteen years, there have been confirmed sightings in most of the states east of the Rockies.

Yet very few people have been fortunate enough to see a Cougar, which makes getting to know her not much easier than getting to know a Yeti. I live right in the epicenter of sightings in Wisconsin, and still I have only seen sign. Cougars are much harder to spot than other predators of similar size, such as Wolves, because they are . . . well . . . catlike. In chapter sixteen, "The Messenger," you saw how hard it could be to even find their tracks. They're solo, nocturnal animals who stick to heavy vegetation, have retractable claws, hardly ever run, and often bury their scat, so they leave little evidence of their presence. Unless, of course, you come across a kill site, as our staff carpenter Mark did.

So why the increase in attacks on humans? I posed the question to Cougar experts of all persuasions, from Don Myers, who has been live-trapping for zoos since 1948, to advocacy groups such as the Mountain Lion Foundation, and I found two predominant theories.

- With Cougars being protected in most Western states, the rebounding population is forcing young males to migrate greater distances to find a niche (like Bears and Deer, female Cougars stay close to their mothers).
- They've always been there, and we are moving into their territory. This is compounded by the fact that we have taken to feeding wild animals, which attracts predators.

The media hoopla over attacks whips up anti-Cougar frenzy. Yet with common sense and some understanding of Cougar's ways, we can peacefully coexist. Here are some tips for honoring our tawny brother that I learned from Cougar experts, research, and my experience with Wolves and other predators:

Out in the Wilds
- Keep a clean camp
- Avoid dead Deer or Elk, especially if fresh or covered with debris/snow—Cougars defend their kills (when we inspected the kill site near our camp, we took the students in as a group)
- Camp or hike with a friend—never alone
- Keep children close by

In Residential Areas
- Don't feed pets outside or leave pets out overnight
- Don't feed wildlife, especially Deer

- Don't go jogging alone—joggers are second to children as attack victims
- Have children play in groups, and indoors by dusk

If You Encounter a Cougar
- Stop—do not run
- Pick up children. Their erratic movements, shrill voices, and tendency to flee say, "Prey!"
- Don't approach, especially if feeding or with cubs, who may be at a kill site when only eight weeks old
- Leave him an exit route
- Back up slowly

If the Cougar Is Aggressive
- Maintain eye contact and keep facing the Cougar
- Stand tall, inflate your size—open coat, arms up, holding branches
- Wave arms, shout, throw rocks, sticks—convince the Cougar you are a threat rather than prey

If Attacked
- Stay on your feet
- Fight aggressively—go for eyes, shove your fist down his throat
- Protect your neck

A zookeeper gave me this advice, "Get your back up against a Tree. You'll look taller and he won't be able to jump on your back—his favorite way to attack."

And above all, use common sense. One of my students told me about his brother-in-law who walked up to a Cougar he saw hiding in a bush. Fortunately, the man wasn't hurt. At the same time, he gave some credence to what a wild animal handler once told me, "Animals are more intelligent than us because they have nothing to prove."

Glossary

Uncommon and tracking-specific terms in the text are defined here.

aboriginal: See **native**.

ancestral memories: The practices and behaviors of our distant Ancestors that proved so essential to their survival and quality of life that they were genetically selected for and transmitted to descendents. Also known as *genetic memory*.

balance: The state of natural order and harmonious functioning of an organism or a system.

become: To assume another identity for the purpose of gaining first-hand, intimate knowledge of the feelings, thoughts, motivations, and circumstance of the entity.

Blackberry Moon: Approximately August. See **moon**.

Budding Leaves Moon: Approximately May. See **moon**.

clan knowledge: A group's collective intelligence that is demonstrated when individuals pool their skills, memories, and reasoning abilities. Enables the group to function better than any member individually.

clan memory: A group's accumulated experiences and teachings that are passed down from generation to generation, usually by Elders through story and example.

ego: The aspect of personality that creates self-consciousness and individual identity. Ego dominance can inhibit development of the tracker-tracked relationship.

envelope: A situation where there are two simultaneously occurring events, where one is not apparent because it is tucked into or otherwise disguised by the other.

envision: To mentally create a scenario for the purpose of discovering its characteristics or outcome without needing to directly observe or experience it, or without having to read its sign.

green season: Summer. Many Northcountry natives traditionally recognize only two seasons: green season and **white season** (winter). Spring and autumn are the transitional times between the two main seasons.

heart-of-hearts: The center of one's being and seat of personal **balance** (see definition), where feelings, intuition, ancestral memories, the senses, and mental input come together to give perspective and guidance.

Hoop of Life: The interrelatedness of an area's plants, animals, and geographical features. An inhabitant's immediate environmental support community. Natives live by the premise that all their needs, whether physical, relational, or spiritual, can be met within their Hoop of Life, i.e., within walking distance. Also known as *Circle of Life*.

hunter-gatherer: See **native**.

moon: The aboriginal equivalent of a calendar month. Encompasses a full twenty-nine-day lunar cycle, which is usually reckoned from new moon to new moon. Commonly named after a significant occurrence during the moon, which in the Lake Superior region could be berry picking (Blueberry Moon, around July) or a change of seasons (Falling-Leaves Moon, around October).

native: A plant or animal living a natural life in her natural habitat. Used synonymously with **hunter-gatherer** and **aboriginal**.

natural: Intrinsic to a species or system.

oneness: A state of being where one is both deeply relaxed and keenly attuned to surroundings. The normal functioning state for most sentient creatures.

Relations: All of the plants and animals who dwell in one's **Hoop of Life** (see definition). Signifies kinship.

shadow: To follow someone and become her mirror image by moving, thinking, and feeling as she does.

shadow track: To **shadow** (see definition) someone without him being present, by listening to the **song of the track** (see definition).

shadow voice: The lingering aftereffects of an animal's call.

shape-shift: To transform into another life form.

Snow-Melting Moon: Approximately April. See **moon**.

Snowshoe-Breaking Moon: Approximately March. See **moon**.

song of the track: The greater voice that guides a tracker; composed of primary sign (track, scat, fur, scrapes, chews), secondary sign (also called *environmental imprint*: altered behavioral patterns, disturbed spiderwebs and vegetation), invisible sign (the affected calls of other animals, **shadow voice** (see definition), **ancestral memories** (see definition), intuitive and instinctive guidance, prior knowledge and experience, and continual questioning.

spirit tracker: A person who can hear the **song of the track** (see definition).

straddle: The average width of an animal's gait, measured from the middle of right footprint to the middle of left. Middle is found by centering a line through print in direction of travel.

Strawberry Moon: Approximately June. See **moon**.

stride: The average length of a single step. Measured from right to left foot, toe to toe or heel to heel, whichever is clearest.

Sun-wise: From east to west; in a circle from east to south to west to north; clockwise.

sweat lodge: A small, domed structure framed with saplings and covered with tarps, skins, bark, or other material. Of American Indian origin and used for sauna-type steam baths, which may be secular, ceremonial, or of a healing nature.

teaching trail: The way to discovery and knowledge that is found by stepping off, or being forced off, the beaten path. Accelerates learning and the revitalizing of one's dormant intuitive tracking ability.

Thunder Beings: In American Indian lore, great Eagle-like Birds who live above the clouds and are respected for their ability to affect the weather, particularly causing thunder, lightning, and rain. Also known as *Thunderbirds*.

turn of the seasons: A year.

white season: Winter. See **green season**.

Wolf-walk: A method of group travel where a leader breaks trail and the others follow in his footsteps. The leader drops back when he tires, to be replaced by a new leader. Used to conserve energy in deep snow conditions, to minimize disturbance in environmentally sensitive areas, and to disguise the number of people in a group.

Index

About the Author

Growing up with woods, fields, and swamps for a backyard, Tamarack Song was usually out the door before breakfast and often in trouble for getting back long after supper. Every time he thought he had a good excuse: a newly discovered pond needed to be explored, nuts or berries had to be gathered, a captivating covey of quail wouldn't let him leave, or a new camp needed just a little more work. And then there was the never-ending issue of late homework—unless it was wildlife art or a nature report.

He had one big fear while growing up—growing up. Most kids older than him were too busy with the modern world to talk with Birds or trail an animal, which edged him deeper and deeper into the natural world. And then one day his father introduced him to a Menominee Indian Elder, who spoke to him in the same wordless language he used with the animals and plants. At that moment he realized that getting older didn't mean you had to give up your wild ways.

Among the experiences that helped shape who Tamarack is today are the years he lived alone in the woods and the time he lived with a pack of semi-domestic Wolves. He has spent his life studying the world's aboriginal peoples, apprenticing to Elders, and learning their survival skills. Of particular interest to him are the language and cultural traditions of the Great Lakes Ojibwe, with whom he has close ties.

To this day, nearly all Tamarack does is related to the hunter-gatherer way of life. He is dedicated to keeping a working knowledge of its practices alive for the coming generations.

You can find out more about Tamarack, the Teaching Drum Outdoor School (which he founded), and his time with Wolves at

www.tamaracksong.org

www.teachingdrum.org

www.brotherwolffoundation.org

About the Artist

Mark Webster is a Tonka Bay–based artist who works in drawing, painting, and sculpture. He received his BFA from the University of Minnesota and specialized in drawing and painting. He has exhibited at the Tweed Museum in Duluth, Minnesota, the Kraine Club Gallery in New York City, and numerous other art venues.

Nature provides him with great inspiration, and he spends many hours in the wild outdoors. His wife, Unsie, is his best friend and supporter.